Jet2.com

NIGEL RICHARDSON

KEY Books

AIRLINES SERIES, VOLUME 26

Front cover image: Jet2.com took delivery of Boeing 737-8MG G-JZBI in February 2018. (Markus Eigenheer, licensed under CC BY-SA 2.0)

Title page image: Departing from Manchester Airport is Boeing 757-236 G-LSAD, which was acquired by Jet2.com in June 2006. (Nigel Richardson)

Contents page image: Boeing 737-8MG G-JZBE of Jet2.com on final approach to Tenerife South Airport, Canary Islands. (Alan Wilson, licensed under CC BY-SA 2.0)

Back cover image: Airbus A321-251NX G-SUNM was delivered to Jet2.com in December 2024. (Richard Jackson)

Published by Key Books
An imprint of Key Publishing Ltd
PO Box 100
Stamford
Lincs PE9 1XQ

www.keypublishing.com

ISBN 978 1 83632 097 5

Typeset by SJmagic DESIGN SERVICES, India.

Contents

Acknowledgements 4

Introduction 5

Chapter 1 Early Years: Express Air Freight and Express Air Services 7

Chapter 2 Channel Express 13

Chapter 3 Jet2.com: The First Eleven Years (2003–13) 37

Chapter 4 Growth and Expansion (2014–23) 68

Chapter 5 Jet2.com in 2024/25 115

Fleet Lists 128

Bibliography 143

Acknowledgements

I would like to thank the following photographers and organisations for kindly allowing the use of their images in the book either directly or through AirHistory.net and Creative Commons: 54north, aeroprints.com, Alec BHX/KKC, Pedro Aragão, Ardfern, Gerry Barron, Ole Johan Beck, Don Bennett, Richard Blata, BriYYZ, Stephen Brown, Shaun Connor, Colin Cooke, Howard J. Curtis, Robert Dowson, Markus Eigenheer, Chris England, Varani Ennio, Laurent Errera, Manuel EstevezR, Derek Ferguson, Ken Fielding, flybyeigenheer, G B_NZ, Dick Gilbert, Don Gilham, Phil Globe, Fergal Goodman, Danny Grew, Thomas Haynes, Gerard Helmer, Graham Hocquard, Rob Hodgkins, Anton Homma, Ian Howat, Richard Jackson, kitmasterbloke, Augusto Laghi, Leeds City Council, LRS747 (Leslie Snelleman), Toni Marimon, Pieter van Marion, Eduard Marmet, David Montgomery, John Murdoch, Till Niermann, Michael Oldfield, Ozzy's Airliners, Arne V Petersen Collection via Jakob L. Dahlgaard, Adrian Pingstone, Riik@mctr, Javier Rodriguez, Eero Saarela, Pertti Sipilä, Kev Slade, John Smitherman, Fred Streep, Björn Strey, Eryk Strzala, Craig Sunter, Barry Swann, Roger Syratt, Bill Teasdale, Trevor Thornton, David Unsworth, Roman Valladares, Peter Vercruijsse, Ronald Vermeulen, Vin LK, John Visanich, Maarten Visser, Alan Wilson, Alec Wilson, Mark Winterbourne and Anna Zvereva. I would like to thank my wife, Gill, for carefully proofreading each chapter of the book. Finally, my thanks to Matt Wharmby and Simona Hill from Key Publishing for advice and support throughout the preparation of the book.

The information and data presented is believed to be accurate up to January 2026.

Introduction

The development and expansion of the inclusive tour package holiday during the 1950s and 1960s led to the establishment of a number of airlines specialising in the provision of charter flights for package holidays (such as Dan-Air). Charter airlines would sell blocks of seats or entire flights to tour operators for integration into package holidays. Over time, the tour operators began to formalise links with holiday charter airlines for their mutual benefit. The tour operators were able to secure seats at reasonable prices for the summer season and could have an input into the scheduling of flights, while the airlines had the assurance of financial security, which could be invested in fleet expansion and modernisation. The process, known as vertical integration, either involved the tour operators owning the airlines or the reverse, where the airline owned the tour operator. Examples include the partnerships between British Eagle Airlines and the Henry Lunn travel agency (later Lunn Poly), Cosmos Tours and Monarch Airlines, the Thomson Travel Group and Britannia Airways (later acquired by TUI), Court Line Aviation and Clarksons Holidays and Intasun Leisure and Air Europe. In some instances, a tour operator established its own in-house airline as part of an integrated travel group, for example the Airtours Group (later renamed MyTravel).

Following the expansion of low-cost airlines, such as easyJet and Ryanair, since the mid-1990s there has been a gradual shift away from traditional package holidays towards independent holidaymaking, with the increasing ability and willingness of consumers to independently source flights, accommodation and other holiday components as part of a 'self' or 'dynamic' package. Some charter airlines responded to the competition from low-cost carriers by introducing and/or expanding their scheduled services, available through online, seat-only bookings.

In February 2003, Jet2.com entered the low-cost airline market. Jet2.com's origin dates back to 1971 and the creation of Carpenter's Air Services, which transported fresh flowers from Guernsey to the UK. From 1975, Express Air Freight was launched to transport fresh vegetables, flowers and other cargo between the Channel Islands and the UK mainland; later expansion led to the creation of Express Air Services, with the introduction of passenger charter flights and acquisition of the first valuable Royal Mail contract.

When the Dart Group purchased Express Air Services in 1983, it was renamed Channel Express. The carrier operated scheduled and ad hoc charter freight services throughout Europe, the Middle East and North Africa as well as passenger charters. The potential afforded by the purchase of two Boeing 737-300(QC) 'Quick Change' aircraft in May 2001, which could be transformed from 17-tonne freighters to 148-seat airliners in less than 40 minutes, led to moderate expansion of its daytime passenger charter operations and was probably a significant influencing factor in the decision to launch Jet2.com in early 2003. It seemed an odd time to enter into such a competitive market with easyJet in the ascendancy and, like Ryanair, having its main base in the UK. Other airlines had already attempted to compete without success. Go Fly (trading as Go), founded as a low-cost subsidiary of British Airways in 1998, was bought by easyJet in 2002. MyTravel Airways launched its scheduled low-cost airline, MyTravel Lite, in October 2002, but by 2005 it was re-integrated into the parent airline. However, Jet2.com's focus was northern England, which it considered to be poorly served by low-cost airlines, and its main strategy was and remains 'Customer First', with the key principles of delivering a friendly service and low-cost flights.

In February 2007, Jet2.com's parent company, the Dart Group, launched Jet2holidays as its tour operator, with a remit to sell package holidays based around the airline's low-cost flight network. Jet2holidays has been highly successful at rejuvenating the inclusive package holiday, and through an integrated structure with its partner airline, it has provided opportunities for Jet2.com to increase the size of both its fleet and its route network, making significant additional contributions to the airline's passenger numbers by including new destinations, thereby reducing financial risk. Today, Jet2.com is the third largest airline in the UK and Jet2holidays is the UK's largest tour operator.

This book describes the history of Jet2.com from its predecessor Express Air Freight, through its transformation into an award-winning scheduled airline and, together with Jet2holidays, forming the Jet2 plc leisure travel group. The book focuses on changes to the airline's aircraft fleet, its route network development and operations up to the present day.

Chapter 1

Early Years: Express Air Freight and Express Air Services

The origin of Express Air Freight (EAF) and Express Air Services (EAS) is linked to the establishment of two businesses by Art Carpenter in 1971: Carpenter's Air Services Ltd and Carpenter's Transport Ltd. Carpenter's Air Services contracted space on cargo airlines to ship fresh flowers and produce from Guernsey to the UK mainland. Initially, two Douglas Dakotas (G-AKNB and G-AMPZ) were chartered from Intra Airways, operating into Bournemouth and East Midlands airports. Carpenter's Transport Ltd arranged for the distribution and delivery of the flowers on the UK mainland to wholesale markets.

Intra Airways, based at Jersey Airport, had been created in January 1969 to operate inter-island air services in the Channel Islands. As an efficient inter-island service in the Channel Islands had also been launched by Aurigny Air Services, Intra focused on passenger and cargo charters from Jersey and Guernsey to the UK mainland and north-west France for the initial three years, using a single Douglas DC-3 Dakota aircraft (G-AKNB). The first Intra freight service took place in March 1970 between Jersey and Bournemouth, transporting Channel Islands flowers and vegetables.

In 1975, Carpenter's Air Services Ltd and Carpenter's Transport Ltd became part of a new Channel Islands freight company, Express Air Freight (CI), with Art Carpenter as Director. The respective roles of the two strands were retained under the umbrella of one organisation. In January 1977, Express Air Freight (EAF), now owned by the Hutton Group through Field Aviation, took over the operation of the two Dakotas chartered from Intra Airways, although they continued to be operated by Intra aircrew. EAF also purchased two Handley Page HPR-7 Dart Herald Series 200 aircraft (G-ATDS and G-BEZB) in August 1977.

Operations using the Heralds, flown by Intra aircrew, to transport flowers and fresh produce from the Channel Islands to Bournemouth Airport and consumer goods from the UK to the Channel Islands began in January 1978. The Guernsey to East Midlands Airport route was operated in association with Air Bridge Carriers, another subsidiary company of Field Aviation, based at East Midlands Airport.

Two more HP Dart Herald Series 200 aircraft (G-BFRJ and G-BFRK) were acquired in April 1978, allowing the introduction of passenger charter flights under the carrier name Express Air Services; the first service was undertaken on 17 March 1978 between Jersey and Bournemouth on behalf of Dan-Air. The two Dakotas operated by EAF were withdrawn from use in September/October 1978, while the Heralds began operating a night-time service for Royal Mail, flying post between London Gatwick, London Stansted and Newcastle.

In late 1978, Intra Airways' Managing Director, Bernard Haddican, was charged with allowing breaches of the Air Navigation Order by encouraging flight crews to exceed their permitted flying hours over a three-year period from January 1976 to November 1978. Nearly 2,000 breaches of aviation law had been committed by Intra Airways during this period and, as a result, Haddican stepped down as Managing Director.

In January 1979, Haddican sold his holding in Intra Airways to the Express Air Freight Group and a major reorganisation took place that resulted in the merger of Express Air Freight and Intra Airways, creating the nucleus of Jersey European Airways. The other majority shareholder, Jersey resident John Habin, who owned Aviation Beaufort (an air charter/taxi business and Piper dealership), wanted Jersey European to establish several key passenger services from Jersey to major airports in the UK. The cargo operation of the new partnership, previously undertaken by Express Air Freight, became part of Express Air Services (EAS).

Douglas DC-3 Dakota G-AKNB was one of the first aircraft to be chartered from Intra Airways by Carpenter's Air Services and then Express Air Freight from 1975. (Chris England)

Douglas DC-3 Dakota G-AMPZ of Intra Airways, which was operated on behalf of Carpenter's Air Services and Express Air Freight. (Roger Syratt)

At the beginning of 1979 Intra Airways had just two Douglas DC-3 Dakotas, one of which was G-AMHJ. (LRS747)

Ownership of Douglas DC-3 Dakota G-AMPY was transferred to Field Aviation, the owner of Express Air Services (EAS)/Express Air Freight (EAF), in January 1980 and it remained in service with the group until February 1982. (Rob Hodgkins, licensed under CC BY-SA 2.0)

Express Air Services acquired Handley Page HPR-7 Dart Herald 209 G-ATDS in August 1977. (Eduard Marmet, licensed under CC BY-SA 3.0)

Handley Page HPR-7 Dart Herald 209 G-BEZB operated for Express Air Freight/Express Air Services between August 1977 and November 1984. (LRS747)

Handley Page HPR-7 Dart Herald 209s G-BFRK, shown here, and G-BFRJ increased Express Air Services' Dart Herald fleet to four aircraft in April 1978. (Dick Gilbert, licensed under CC BY-SA 4.0)

At the beginning of 1979, Intra had three Viscounts (G-AVJB; G-BAPE; G-BAPG) but its Dakota fleet had been reduced to just two aircraft (G-AMHJ and G-AMPY), of which only one retained the 36-seat passenger configuration. Before Jersey European became fully operational, Intra continued to use the Viscounts on summer season scheduled flights and weekend charters linking the Channel Islands with Amsterdam, Brussels, Cambridge, Carlisle, Edinburgh, Newcastle, Ostend, Swansea and Bournemouth (the latter for Dan-Air). Services from Birmingham and Gatwick were also flown by the Viscounts, together with a significant number of ad-hoc charters to locations in Europe.

Jersey European Airways began trading on 1 November 1979 under the chairmanship of John Habin. Under the new arrangements, some of Intra's scheduled routes using the inherited Vickers Viscount and Dart Herald aircraft were retained, including the Jersey–Cambridge passenger service and the Channel Islands–Bournemouth cargo link. Express Air Services secured a Royal Mail delivery contract between Bournemouth, Bristol and Liverpool in 1980, on which it operated Dakotas and Dart Heralds.

In October 1980, Express Air Services withdrew from the partnership with Jersey European to resume operations under its own title. It took with it the four Dart Heralds and two Dakotas; the Viscounts were not used by EAS. G-AVJB was transferred to Jersey European while G-BAPE and G-BAPG were acquired by Southern International Air Transport. From its main base at Bournemouth Airport, Express Air Services operated up to 20 weekly scheduled cargo return flights, alongside Royal Mail contract flights at night and occasional cargo services transporting car parts to assembly plants in Europe. The end of service for the two Dakotas with EAS finally arrived in February 1982 when they were sold to Air Atlantique. In June of the same year, a de Havilland Canada DHC-6 Twin Otter (G-RBLA) was acquired to operate a small number of passenger charters.

Chapter 2

Channel Express

In 1983, Express Air Services and associated companies were acquired by the Phil Meeson Group and renamed Channel Express. It marked the beginning of a period of expansion. Although two Dart Heralds (G-BFRJ and G-BFRK) were sold in 1984, a third Dart Herald (HPR-7 Dart Herald 214 G-BEBB; re-registered as G-CEAS in January 1986) was bought from British Air Ferries in April 1985 to operate a newly gained overnight, time-sensitive freight service between Birmingham, Nuremberg and Hamburg on behalf of TNT-IPEC.

A new contract from FedEx involving an overnight freight service between London Stansted, Birmingham and Brussels was secured in 1986, which led to the return of two of Channel Express's former Dart Heralds, G-BFRK and G-BFRJ, re-registered as G-GNSY and G-CEXP respectively. More freight contracts were obtained in 1987 from the Royal Mail, Guernsey Post Office and several major newspaper publishers, as well as ad-hoc passenger/freight sub-charters from various organisations. This led to further expansion of the Dart Herald fleet, which by August 1988 had increased to eight aircraft.

In 1988, the Channel Express Group was floated on the UK Unlisted Securities Market (USM) to generate additional funds for further expansion. Over 11 million ordinary shares were issued following the flotation. The anticipated expansion continued, with the agreement of more Royal Mail contracts and a freight contract from United Parcel Service (UPS).

Fleet growth and enhancement continued in March 1989 when Channel Express sub-leased a Lockheed L-188A(F) Electra freighter from Air Bridge Carriers. The Electra, powered by four wing-mounted Allison 501-D13 turboprop engines, was capable of carrying a payload of 15 tonnes, more than double that of the Dart Herald. It provided a maximum volume of 95 sq m to carry eight 108 x 88in (2.74 x 2.24m) or six 125 x 88in (3.18 x 2.24m) pallets or freight containers. Channel Express was so impressed with the relatively fast loading/unloading times of the Electra and its capability for transporting pallets, containers and bulk loads that the company purchased its own example (L-188C(F) Electra N347HA) in November 1989. The Electras were introduced on the high-density Channel Island routes and were frequently used to fly car components for Ford from various UK airports to Saarbrücken, Turin and

Valencia. They also operated express package flights between Southend and Cologne for UPS.

Another new aircraft type was introduced in February 1990 when Channel Express acquired a Fokker F27-600 Friendship (G-FEAE). The Fokker Friendship was a high-wing aircraft, fitted with a pair of Rolls-Royce Dart turboprop engines. The F27-600 variant was a quick change cargo/passenger version with a large cargo door, and this one was used to operate an overnight service from Birmingham to Brussels on behalf of FedEx. The Dart Heralds were also used on an overnight service for FedEx, operating between Prestwick, Manchester and Brussels from November 1990.

Benair Freight Ltd was acquired by the Philip Meeson Group in June 1990 after an additional one million shares were issued to fund the investment. Benair was an international freight-forwarding management company with well-established links to the Far East through specialisation in the trade of tropical fish. It provided Channel Express with an international presence, having branches and offices in Singapore, Hong Kong, Kuala Lumpur and Taiwan.

Handley Page HPR-7 Dart Herald 214 G-CEAS was acquired by Channel Express from British Air Ferries in April 1985, registered as G-BEBB. It was re-registered as G-CEAS in January 1986. (Bill Teasdale)

Handley Page HPR-7 Dart Herald 209 G-GNSY was converted to a freighter and returned to service with Channel Express in June 1987, having previously operated with the carrier until October 1984 as G-BFRK. (G B_NZ, licensed under CC BY-SA 2.0)

Handley Page HPR-7 Dart Herald 209 G-CEXP was another Dart Herald that returned to Channel Express after being converted to a freighter (previously operated as G-BFRJ). (Shaun Connor)

Channel Express operated Handley Page HPR-7 Dart Herald 210(F) G-SCTT between August 1988 and February 1997. (Colin Cooke, licensed under CC BY-SA 2.0)

Handley Page HPR-7 Dart Herald 401(F) G-BEYF of Channel Express at London Stansted Airport in April 1995. (LRS747)

In November 1989, Channel Express bought its first Lockheed L-188C(F) Electra, registered as N347HA, due to the type's capability for transporting pallets, containers and bulk loads and its relatively fast loading/unloading times. (Fergal Goodman)

In August 1990, the first of Channel Express's re-engineered and upgraded Dart Heralds was rolled out. The aircraft, dubbed the 'Super Herald', included improved instrumentation and avionics and an increase in the payload carried to six tonnes. The improvements were completed on several more aircraft to extend their lifespan.

The Lockheed Electra fleet was increased to four aircraft in the autumn of 1990 following the acquisition of L-188A(F) Electras N5535 and N343HA and L-188C(F) Electra N5539, mainly to operate the expansion of UPS services, which now included Helsinki (Finland), Stockholm (Sweden) and Bergamo in northern Italy. Stage 3 noise certification of the Electra was completed in 1991 to allow the type to operate into more noise-conscious airports of northern Europe. A Dart Herald was also assigned to a UPS freight contract between Shannon (Ireland) and Paris Orly Airport.

On 31 May 1991, the company name was changed from the Channel Express Group plc to the Dart Group plc after the parent company was floated on the London Stock Exchange. Another share issue of 2.75 million ordinary shares was used to raise additional funding for expansion, bringing the total number of ordinary shares in the company to 15.25 million. The Dart Group comprised an aviation services division, provided by Channel Express (Air Services) and

Deltec Aviation Services (the latter being responsible for the supply and repair of aircraft instrumentation and radio equipment), and freight forwarding and distribution divisions, including Benair Freight Ltd, Channel Express Ltd and Channel Express (CI) Ltd.

The withdrawal of FedEx from European operations in 1991 resulted in Channel Express losing some valuable contracts. However, the potential detrimental impact was allayed by an expansion of operations for Royal Mail as part of its domestic overnight airmail network, Skynet, which involved trunk and feeder services to many UK regional airports. A Lockheed Electra was assigned to this first containerised freight operation, which began in 1992 between London Stansted and Edinburgh and carried up to 4.3 million letters annually, generating £2.8 million revenue for the carrier.

Three more Dart Heralds were overhauled, upgraded and refurbished during 1991 and early 1992. The Dart Herald aircraft based at Bournemouth continued to operate scheduled services to and from the Channel Islands, transporting newspapers, general freight and flowers. Two Heralds were also operated at night between Coventry and Edinburgh on a contract for Parcelforce, the route being eventually taken over by the Electras as the volume of cargo increased. From mid-1993, several Heralds began operating Royal Mail flights between Newcastle and Cardiff following the collapse of Celtic Airways; operations involving the Electras were expanded through regular freight flights on behalf of Finnair.

The sale of Southend Airport to Regional Airports Ltd in August 1993 resulted in the closure of UPS's Southend base and brought an end to Channel Express's Southend to Cologne service. However, UPS opened a new facility at East Midlands Airport, which led to the Electras being redeployed on an express package service between Shannon/Dublin, East Midlands and Cologne (Germany).

During 1994, Channel Express took the decision to build up a fleet of Fokker F27 Friendships, having operated its first example between February 1990 and May 1992. This was in preparation for the eventual replacement of its ageing Dart Herald fleet; aircraft were retired as major maintenance checks became due. The F27 was a more modern aircraft with a better performance and a larger cargo deck volume (64 sq m) and payload (6.5 tonnes) compared to the Herald. It could operate over a range of 1,000nm and was able to meet Stage 3 noise standards. It opened up opportunities for Channel Express to increase business in the cargo charter market throughout Europe. The second F27 Friendship to be operated by Channel Express, an F27-600 variant, was purchased in November 1994 and registered G-BNIZ, followed by two more, F27-600 G-CHNL and F27-500 G-CEXB, in 1995 and F27-500 G-CEXA in January 1996. During 1995, the F27s operated a six-month freight contract for Alitalia and the Italian Post Office, flying between Genoa, Corsica, Sicily and Rome, and operated a similar service for the Swedish Post Office. They were also used for urgent ad-hoc freight charters.

Lockheed L-188A(F) Electra N343HA was leased by Channel Express from Zantop International Airlines for three periods between November 1990 and April 1997. (Arne V. Petersen Collection via Jakob L. Dahlgaard)

Lockheed L-188A(F) Electra G-CHNX was initially acquired by Channel Express as N5535 in October 1990. Following a period on lease to Iona National Airways from October 1993 to November 1994, it returned to Channel Express and was re-registered G-CHNX. It remained in service with the carrier until October 2001. (Gerry Barron)

Another Lockheed L-188A(F) Electra leased by Channel Express from Zantop International Airlines due to the expansion of its freight services was N341HA. (Gerard Helmer)

Lockheed L-188C(F) Electra N360Q departing From Dublin Airport in June 1999 while operating for Channel Express on lease from Renown Aviation. (Fergal Goodman)

Channel Express began operating a Fokker F27 Friendship in February 1990. In 1994, the carrier decided to build up a fleet of F27 Friendships and purchased its first example, F27-600 Friendship G-BNIZ, in November 1994. (aeroprints.com, licensed under CC BY-SA 3.0)

Fokker F27-600 Friendship G-CHNL was operated by Channel Express from March 1995 to 12 January 1999, when it was destroyed following a crash during the final stages of approach to Guernsey Airport. (Augusto Laghi)

Fokker F27-500 Friendship G-CEXB was acquired by Channel Express in November 1995 and remained with the carrier until March 2005. (aeroprints.com, licensed under CC BY-SA 3.0)

By early 1996, Channel Express had a fleet of four Fokker F27 Friendships after F27-500 Friendship G-CEXA was acquired in January of that year. (Anton Homma)

At the end of the 1994/95 financial year, the Dart Group had made a pre-tax profit of £2.96 million on a turnover of £58.2 million, representing an 11.7 per cent improvement on the previous year. The financial progress was partly attributable to the acquisition of Fowler Welch Ltd, a temperature-controlled produce distribution company, at the end of October 1994. The company specialised in the distribution of home-grown and imported produce from its facilities in Lincolnshire and directly complimented the other part of the Distribution Division, Channel Express Ltd, which undertook the distribution of flowers and produce, including goods from the Channel Islands through Channel Express (CI).

In anticipation of an expected growth in air freight transport and an increasing expectation from customers, on cost grounds, for cargo carriers to operate larger aircraft with greater capacity rather than use several smaller aircraft to carry the same load, Channel Express purchased an Airbus A300B4-103 in February 1996; the carrier's first jet aircraft. The Airbus A300B4 is a wide-body medium- to long-range airliner powered by two General Electric CF6-50A engines. It was the first aircraft type to be developed and produced by Airbus. As a passenger-carrying airliner it had 280–300 seats in a typical configuration and could operate up to a range of 2,550nm. The aircraft acquired by Channel Express was previously operated by Eastern Air Lines and VIASA of Venezuela. Prior to entering into service with Channel Express, it needed to be converted from a passenger aircraft to a freighter. The conversion was carried out by British Aerospace Aviation Services at Filton, Bristol and was the first such A300 conversion undertaken by the company. The work involved installing a large main deck cargo door and fitting a strengthened floor in the main cabin for transporting large and heavy cargo. On completion, the aircraft had a maximum payload capability of 45 tonnes of palletised cargo and more than 350 cubic metres volume, with up to 20 main deck positions. The under-floor hold could accommodate up to 20 standard-size ULD containers. The converted A300 was named 'Eurofreighter' by Channel Express. After gaining Supplemental Type Certification for the conversion configuration from the UK's Civil Aviation Authority (CAA) and the Federal Aviation Administration (FAA) of the USA, the aircraft was registered G-CEXC and entered service on 1 August 1997. It began by flying daytime weekday services from Sweden to Luxembourg on behalf of the Swiss freight forwarding and logistics company Panalpina, an express parcel service for TNT at night and operated on behalf of British Airways World Cargo from London Stansted to Tel Aviv at the weekend.

Four F27s, five Electras and seven Dart Heralds continued to operate throughout Europe for Channel Express during 1996 on behalf of express parcel carriers, postal services and freight forwarders, as well as flying freight services between the UK mainland and the Channel Islands to transport produce, newspapers and mail. In March and December 1996, Channel Express retired two of its Dart Heralds (G-CEXP and G-GNSY) and by August 1997 all but one of the Heralds (G-BEYF) had been withdrawn from service. As a replacement, three F27s were purchased in early 1997, one F27-600 from Fokker Services (G-CEXD) and two F27-500s from Egyptair (G-CEXE and G-CEXF), forming a fleet of seven aircraft. Two Electras (N341HA and N343HA) were returned to the lessor (Zantop International Airlines) in April 1997, although another aircraft (L-188C(F) Electra N360Q) was leased to ensure customer contracts could be fulfilled.

Channel Express introduced its first jet aircraft into service in August 1997, the Airbus A300B4. The converted wide-body passenger aircraft provided much greater capacity than the smaller aircraft in the fleet. Airbus A300B4-103(F) G-CEXC, shown here, was initially acquired by Channel Express in February 1996 and converted from a passenger aircraft to a freighter by British Aerospace before entry into service in 1997. (Danny Grew)

Handley Page HPR-7 Dart Herald 213 G-AVPN was one of the last Dart Heralds to be operated by Channel Express. It was acquired in January 1992 and remained in service with the carrier until July 1997. On retirement it was transferred to the Yorkshire Air Museum for preservation; however, it was reduced to just the cockpit section in 2015 due to corrosion. (Shaun Connor)

In 1997, Channel Express purchased three Fokker F27 Friendships to replace Dart Heralds which, with the exception of one aircraft, had been withdrawn from service. Fokker F27-500 Friendship G-CEXE joined Channel Express in April 1997 and was the last example of the type to be withdrawn from use by the carrier. (Don Gilham)

Fokker F27-500 Friendship G-CEXF was operated by Channel Express between April 1997 and October 2004 before moving to Turkish carrier MNG Airlines. (Peter Vercruijsse)

Fokker F27-500 Friendship G-CEXG was the ninth and final F-27 Friendship to join Channel Express. It was acquired from Jersey European as G-JEAP in October 2000 and re-registered as G-CEXG the following month. The aircraft remained in service with Channel Express until March 2005. (aeroprints.com, licensed under CC BY-SA 3.0)

Channel Express purchased two more A300B4s in July 1997 and February 1998; both were ex-Continental Airlines aircraft. They were the A300B4-203 variant, which had an increased maximum take-off weight of 165 tonnes (364,000lb) compared to the 158 tonnes (348,330lb) of the -103 variant. Due to other commitments, British Aerospace Aviation Services was unable to complete the conversion of these aircraft so the work was undertaken by Daimler-Benz Aerospace Airbus (DASA) in Dresden, Germany. DASA had previous experience of completing passenger-to-freighter conversions of Airbus A310s and several A300s. The converted aircraft entered service in April (G-CEXH) and September 1998 (G-CEXI), operating express parcel services for TNT and then a long-haul Stansted–Cologne–Dubai route on behalf of UPS. With the growth of air freight, the three Eurofreighters were operating at full potential and, following an increase in demand for use of the type, Channel Express acquired another A300B4-203(F) in late 1998, on lease from the Finova Capital Corporation. The former Malaysian Airlines and Air Maldives airliner had been converted to a freighter before beginning operations with Channel Express in March 2000 as G-CEXJ. Several A300s were also acquired in 1999 for use as a source of spare parts for Channel Express's A300 freighters, which led to the creation of Channel Express Parts Trading, a supplier of spare parts for other Airbus users.

Between October 1999 and July 2000, Channel Express operated a TNT Express A300B4-203(F) (G-TNTI), on behalf of the owner. Channel Express's first Eurofreighter, G-CEXC, was also sold to TNT Express in February 2000

The second Airbus A300B4 freighter to enter service with Channel Express was a -203 variant, G-CEXH, in March 1998. The -203 had a higher maximum take-off weight than the -103 variant. (Ole Johan Beck)

Airbus A300-B4-203(F) G-CEXI of Channel Express being loaded at Helsinki-Vantaa Airport, Finland, in March 2003. (Pertti Sipilä)

A former Malaysian Airlines and Air Maldives airliner Airbus A300-B4-203 was acquired by Channel Express in late 1998 and converted to a freighter before entry into service as G-CEXJ in March 2000. (Howard J. Curtis)

(re-registered as G-TNTS) but continued to be operated by Channel Express for TNT until October 2000.

At the start of the new millennium, Channel Express had a fleet of 13 aircraft, including three A300B4-203(F)s, four Lockheed Electras and six Fokker F27 Friendships. The Dart Group was developing into a leading organisation for freight transport, distribution and delivery throughout the UK and Europe, and to the Middle East. An issue of 1.622 million ordinary shares in September 1999 was used by the Group to generate funds to continue to develop the Distribution Division through the purchase of Coolchain, a company which distributed locally grown and imported fruit from its centres in Kent.

In May 2001, Channel Express acquired two ex-Lufthansa Boeing 737-330(QC) aircraft (TF-ELP and TF-ELR) from Icelandic airline Islandsflug in preparation for entering the inclusive tour charter airline business market. The Boeing 737-300 variant first flew in February 1984 and went into service the following December. It was a member of the second generation or 'Classic Series' of the Boeing 737. The QC or 'Quick Change' variant of the aircraft could be converted from a 148-seat airliner to a freighter, capable of carrying a 17-tonne payload, within 40 minutes, which offered the opportunity of maximising the use of the aircraft. One of the 737(QC)s was based at Stansted Airport and the other at Edinburgh Airport, from where they operated the high-volume overnight mail service on behalf of Royal Mail. During the daytime, they were involved with flying passenger charters for tour operators, having been contracted to operate approximately 1,000 hours per aircraft for the May to October summer season.

Four more Boeing 737-300s were acquired during 2002. They were all ex-Ansett Australia aircraft, purchased for a relatively low price after the Australian airline had gone into administration in March 2002. Three of them remained as passenger aircraft (G-CELS, G-CELU and G-CELV), and the fourth (G-CELW) was converted into a Special Freighter (SF) before being re-delivered to Channel Express in October 2003.

A fourth Eurofreighter (G-CEXK) also joined the fleet in April 2002. The former Farnair Europe A300B4-103(F), which was leased from the Finova Capital Corporation, had already been converted to a freighter. The Eurofreighters were kept busy operating for two major European express parcel delivery organisations, flying from the UK, Ireland and Italy to freight hubs in Belgium and Germany.

In October 2002, the Dart Group announced that it planned to enter the low-cost airline market and launched Jet2.com as a separate trading name of Channel Express (Air Services). A lengthy period of research into the low-cost airline industry and use of a population census model for identifying potential markets was undertaken in order to inform the selection of a base for the airline. Leeds Bradford Airport emerged as the strongest option.

In May 2001, Channel Express acquired two Boeing 737-330(QC) aircraft from Icelandic airline Islandsflug, including TF-ELP pictured here, in preparation for entering the inclusive tour charter airline market. (Ken Fielding, https://www.flickr.com/photos/kenfielding, licensed under CC BY-SA 3.0)

Boeing 737-330(QC) TF-ELR of Channel Express at Faro Airport, Portugal, in July 2002. (Pedro Aragão, licensed under CC BY-SA 3.0)

A fourth 'Eurofreighter' joined the Channel Express fleet in April 2002 following the arrival of an ex-Farnair Europe Airbus A300B4-103(F) on lease from the FINOVA Capital Corporation, registered as G-CEXK. (Eero Saarela)

Channel Express acquired Boeing 737-377 VH-CZG from Ansett Australia in August 2002. It was converted to a Special Freighter (SF) before entering into service as G-CELW in October 2003. (Howard J. Curtis)

Boeing 737-377 G-CELS was acquired by Channel Express in May 2002 and initially had a two-month period on lease with Polish airline Buzz before returning in September 2002 to operate for Channel Express and Jet2.com. (Leeds City Council)

Another ex-Ansett Australia Boeing 737-377, G-CELU, is pictured taxiing out for departure at Manchester Airport. (Nigel Richardson)

Boeing 737-377 G-CELV joined Channel Express in October 2002 and began passenger services for Jet2.com in February 2003. (Nigel Richardson)

Following the Dart Group's decision to develop and increase its operations, two more ex-Ansett Australia 737-300s were purchased during 2002 and delivered in 2003. An agreement was made with Ansett Australia's receivers in June 2003 to purchase eight additional 737-300s, with delivery of six of them before the end of the year, to create a fleet of 14 aircraft. By October/November 2003, 737-330(QC)s TF-ELP and TF-ELR had been placed on the UK register as G-CELP and G-CELR respectively.

Four of the 12 737-300s purchased in 2002–03 (G-CELA, G-CELX, G-CELY and G-CELZ) were converted to Quick Change configuration. The installation of a large front cargo door and a roller flooring system allowed the seats, which were mounted on a number of carpeted pallets, to be transferred to and from the aircraft through the cargo door using standard or specialised cargo-loading equipment. The four aircraft joined the two previously purchased 737-300(QC)s (G-CELP and G-CELR) either flying night-time freight services for Royal Mail or as overnight express carriers and passenger charters during the day. The Royal Mail aircraft were based at Belfast, Edinburgh, Exeter, Leeds Bradford and Stansted airports. The Leeds Bradford and Belfast aircraft supported the operations of Jet2.com, while the Stansted and Edinburgh aircraft operated scheduled and charter passenger services for other airlines and only occasionally for Jet2.com. The other seven unmodified 737-300s were operating for Jet2.com by 2004.

A consequence of the introduction and growth of the Boeing 737 fleet from 2001 was the gradual withdrawal from service of the Lockheed Electras. The active Electra fleet was reduced to just two aircraft by the end of 2001 and the final aircraft, G-CEXS, was retired in May 2003. The Fokker F27s had continued to operate contracted services on behalf of Royal Mail and newspaper publishers as well as scheduled freight services to and from the Channel Islands. However, the requirement for containerised jet aircraft to be used for the Royal Mail contract and the increased competition for freight services to the Channel Islands resulted in the gradual withdrawal from service of the F27s. Three aircraft (G-CEXA, G-CEXF, G-CEXD) had been retired by autumn 2004 and the final four (G-CEXB, G-CEXE, G-CEXG, G-BNIZ) were withdrawn in early 2005.

During 2005, UPS revised its aircraft requirements as part of the ongoing development of its European business. As a consequence, Channel Express was given notice of the termination of the European freight services that it had been operating for UPS since 1988; hence, it was decided to cease operating the A300B4(F)s. Both of the leased A300B4(F)s (G-CEXJ and G-CEXK) were withdrawn from use in October 2005 and returned to the lessor in early 2006. The Group's two owned A300B4(F)s (G-CEXH and G-CEXI) were sold. Channel Express operated its final A300B4(F) flight between Milan Bergamo Airport and Cologne for UPS on 3 February 2006.

In August 2005, the Dart Group transferred its shares from the main listing on the London Stock Exchange to the Alternative Investment Market (AIM), a sub-market of the Stock Exchange for smaller companies, which had replaced the Unlisted Securities Market. The less regulated market was considered to be more suited to the entrepreneurial and more speculative business approach of the Group. The Company also divided each ordinary share by four to make them more marketable, leading to 134 million shares in issue. In the same month, the Dart Group sold Benair Freight International, its freight forwarding business, and proceeded to sell its Channel Islands distribution company, Channel Express (CI) Ltd, the following year. This left the Distribution division comprising Fowler Welch-Coolchain, a UK specialist temperature-controlled distribution company, and the recently acquired R. F. Fielding Cheshire Ltd, a non-temperature-controlled distribution business. From January 2006, the Dart Group's aviation activities, namely Channel Express (Air Services) and Jet2.com, were harmonised under the Jet2.com Air Operators Certificate and Channel Express (Air Services) Ltd changed its name to Jet2.com Ltd.

Boeing 737-330(QC) TF-ELR was placed on the UK register as G-CELR in October 2003. It was on lease to Flyglobespan from Channel Express/Jet2.com between March 2003 and May 2007, and is shown here in a hybrid Jet2.com livery. (Björn Strey, licensed under CC BY-SA 2.0)

In 2003, Channel Express/Jet2.com took delivery of four Boeing 737-300s which were converted to a Quick Change (QC) configuration before entering service, including 737-377 G-CELA pictured here at Manchester Airport. (Nigel Richardson)

Ex-Ansett Australia and Qantas Boeing 737-377 VH-CZB joined Channel Express as G-CELX in March 2003 and began operating for Jet2.com the following month. (Nigel Richardson)

Boeing 737-377(QC) G-CELY operating for Jet2.com at Belfast International Airport. (Ardfern, licensed under CC BY-SA 3.0)

The final Boeing 737-377 to be delivered to Channel Express from Ansett Australia during 2003 was registered G-CELZ. It was converted to a QC configuration and entered into service with Jet2.com in April 2004. (Derek Ferguson)

Chapter 3

Jet2.com: The First Eleven Years (2003–13)

In October 2002, the Dart Group announced the creation of Jet2, a new low-cost, no frills airline within the Channel Express (Air Services) division. Although the existing freight and passenger charter operations were expanding and proving successful, there was a dependence on a relatively small number of customers. The Group considered that by using its existing resources and infrastructure, the introduction of Jet2 would enhance the potential for further growth within the division, especially in the light of the recent success of other low-cost carriers such as easyJet and Ryanair. The company stated that its philosophy was 'to offer a friendly service at the lowest possible cost, thereby maximising the opportunities for our customers to travel'.

Prior to the announcement of the new airline, the Group had explored the demographics of the population catchment areas around potential bases throughout the country. Bournemouth Airport was short-listed as it was the home of Channel Express's administrative headquarters, but Leeds Bradford Airport was eventually identified as the option with the greatest potential. The region was generally poorly served in terms of low-cost airlines, and this often resulted in customers having to take relatively long journeys to other airports, including East Midlands, Manchester, Liverpool, Luton or London Stansted for low-cost air travel, despite the fact that approximately 9.5 million people lived within a 90-minute drive of Leeds Bradford. The airport was also very supportive, being very keen to accommodate a low-cost airline to serve the region.

Jet2, under the brand name Jet2.com, began operating as an airline with its first passenger flight between Leeds Bradford and Amsterdam on 12 February 2003 as part of a twice-daily weekday service. From 1 May 2003, more services were added, including daily flights to Barcelona, Malaga, Milan and Palma, flights to Nice three times per week and flights to Alicante six times per week. Flight bookings were made either online through the Jet2.com website or through a call centre. The airline initially used three Boeing 737-377s (G-CELS, G-CELU and G-CELV) and in July four more aircraft were acquired: 737-377 G-CELB and 737-33As G-CELC, G-CELD and G-CELE. The number of 737-300(QC) aircraft had also increased to five by September 2003 and several of these were used to support Jet2.com's operations when they weren't involved with freight activities for Channel Express.

The overall growth of the fleet led to the addition of new services to Prague in September and to Geneva for the winter season. However, there was a reduction in the frequency of flights to most destinations during the winter months and services to Milan were discontinued. The public response to Jet2.com was very positive. It carried its 100,000th passenger in June 2003 and, during that year, Jet2.com and Channel Express (through its separate charter operations for other airlines) combined carried 604,563 passengers, an increase of 900 per cent compared to the 67,180 carried in 2002.

For the 2004 summer season, Jet2.com added new services to its existing scheduled flights from Leeds Bradford, including daily or twice-daily flights to Belfast, Paris and Murcia and flights four times per week to Faro and Venice. At the end of April 2004, Belfast became a second base, with the introduction of a service to Prague, followed later in the year by a service to Barcelona, both involving daytime flights to maximise use of the Boeing 737-300(QC) aircraft based at the airport. The demand for flights from both Leeds Bradford and Belfast during the summer season exceeded Jet2.com's expectations. More than 500,000 passengers were flown from the UK to the Mediterranean and city break destinations between 1 April and 30 September 2004, culminating in one million passengers carried by Jet2.com since its inception. In order to keep pace with its expansion plans, Jet2.com acquired six more Boeing 737-300s between August 2004 and February 2005. Two ex-Lufthansa 737-330s arrived in August and September, registered as G-CELH and G-CELI, followed by another two, ex-Ansett Airlines 737-377s (G-CELF and G-CELG) in October. Two 737-330s, previously operated by Lufthansa and Bulgarian Airlines/Bulgaria Air, joined the Jet2.com fleet as G-CELJ and G-CELK in December 2004 and February 2005 respectively.

In December 2004, Jet2.com opened its third base at Manchester Airport. Seven 737-300s were based there, operating flights to UK domestic, Mediterranean, European city break and winter ski destinations, including daily services to Alicante, Faro, Geneva (during the winter season only), Malaga, Murcia, Nice, Pisa, Valencia and Venice, and two- or three-times daily flights to Amsterdam and Edinburgh. A three-times daily Manchester–London Gatwick service was introduced in January 2005. As another example of maximising the use of aircraft, the Manchester–Edinburgh service was operated by the 737-300(QC)s based at Edinburgh.

Jet2.com began to diversify and modernise its fleet in 2005. In May, its parent company, the Dart Group, announced the acquisition of two Boeing 757-200s, a new and larger aircraft type. Boeing had begun to develop the 757 during the 1970s as it looked to replace the Boeing 727-200. The 757 is the product of cutting-edge design concepts and refinements. It features an advanced technology wing design, use of a large amount of advanced composites and other improved lightweight materials

Boeing 737-33A G-CELC of Jet2.com at Palma Airport, Mallorca, Spain, in January 2004. (Toni Marimon)

Painted in 'Jet2 Yorkshire' special colours is ex-Ansett Australia Boeing 737-377 G-CELB. (Ronald Vermeulen)

Boeing 737-33A G-CELD of Jet2.com on final approach to Barcelona-El Prat Airport, Spain, in May 2009. (Laurent Errera, licensed under CC BY-SA 2.0)

Boeing 737-33A G-CELE of Jet2.com first entered commercial service with Monarch Airlines in October 1988 as G-MONN, going on to operate for Euroberlin and Ansett Australia (VH-CZX) before becoming part of Jet2.com's fleet in July 2003. (Pedro Aragão, licensed under CC BY-SA 3.0)

Ex-Lufthansa Boeing 737-330 G-CELH was acquired by Jet2.com in August 2004. (Nigel Richardson)

Jet2.com's Boeing 737-330 G-CELI painted in 'Jet2 Manchester' special colours pictured here at Manchester Airport. (Nigel Richardson)

Delivered to Jet2.com in October 2004, G-CELF was another ex-Ansett Australia Boeing 737-377. (Nigel Richardson)

Boeing 737-377 G-CELG was operated by Jet2.com from October 2004 until it was retired in April 2018. (Nigel Richardson)

Previously of Lufthansa and Bulgarian Airlines, Boeing 737-330 G-CELJ of Jet2.com is taxiing to a runway exit following arrival at Manchester Airport. (Nigel Richardson)

Boeing 737-330 G-CELK was operated by Jet2.com from February 2005 until October 2017. (Nigel Richardson)

in the construction of the airframe, and second-generation, high by-pass turbofan engines, all of which contribute to its exceptional performance characteristics and low operating costs. The 757 prototype made its maiden flight in February 1982 and commercial service began in January 1983. The 757-200 offered Jet2.com a higher seating capacity (up to 235 seats) and a greater range (up to 3,500nm) than the Boeing 737-300, providing the opportunity for the carrier to expand its network to holiday destinations in the Canary Islands and the Eastern Mediterranean region. The first Boeing 757 (757-236, G-LSAA) was acquired by Channel Express in May 2005 and, after being fitted out with a high-density seating configuration, entered service with Jet2.com in October 2005. It was just over 17 years old and had previously been operated by Air Europe, Air Europa, Iberia, Air Anatolia and Fly Air. The second example (757-27B, G-LSAB), also acquired by Channel Express in May 2005, was of similar age and began commercial services for Jet2.com in December 2005 after previously plying its trade with Air Holland, Odyssey International, Sterling Airways, Britannia Airways, El Al Israel Airlines, Martinair, Air Anatolia and Fly Air. Both aircraft were initially used on the new Leeds Bradford to Tenerife route, which began in October, and on winter services to Faro (Portugal) and ski destinations including Geneva and Chambery in south-east France.

During the 2005 summer season, Ibiza was added to the route network from Leeds Bradford and the frequency of flights to the popular holiday destinations of Malaga, Alicante, Murcia and Palma were increased to either two or three times daily. A daily flight from Belfast to Bournemouth was included in the summer schedule and 14 destinations were served from Manchester. The frequency of flights to Geneva from Leeds Bradford for the winter ski season was increased to eight per week.

Jet2.com opened two new bases in the second half of 2005 – Newcastle International Airport and Blackpool Airport – making a total of five bases in the northern UK. Operations began at Newcastle on 4 September, initially to Murcia, with Cork added at a later date. The first service from Blackpool was to Belfast, beginning on 4 November, with services to Murcia and Palma introduced later in the year and to Alicante from April 2006. Edinburgh Airport became a sixth base in April 2006.

During 2005, Jet2.com carried 2.377 million passengers at a load factor of 74.5 per cent, having flown 24,150 stage flights. In addition to serving Jet2.com's programme of scheduled flights, additional use of the 737 fleet was achieved through charter operations on behalf of third parties, including other airlines, tour operators and sports groups. The Dart Group reported a profit before tax of £15.0 million for the 2005/06 financial year, an increase of 78.6 per cent compared to the previous year (£8.4 million), and a turnover of £319.6 million (2004/05: £268.0 million).

At the beginning of 2006, Channel Express (Air Services) Ltd changed its name to Jet2.com Ltd. Jet2.com had a fleet of 21 Boeing 737-300s, of which six were Quick Change variants based at Belfast, Edinburgh, Exeter, Newcastle and Stansted (enabling them to operate night services for Royal Mail), and two Boeing 757-200s. By the end of the year, the 757 fleet had increased to eight aircraft. 757-23A G-LSAC, an ex-Air Transat and Ryan International Airlines aircraft, was acquired by the Dart Group in March 2006 and introduced into service the following May. The next one to enter service was 757-236 G-LSAD in June 2006, having previously been operated by Air Europe and Greece Airways. A former Air Holland and Air Slovakia 757-27B was acquired in June 2006 and registered as G-LSAE, and three ex-China Southern Airlines 757-21Bs arrived in November and entered service as G-LSAG, G-LSAH and G-LSAI. One aircraft, registered as G-LSAF, was acquired on lease from GATX in October 2006. However, it was returned to the lessor in January 2007 without ever entering service.

An increased range of destinations was introduced to the summer 2006 schedules, including Menorca, Milan, Pisa, Rome and Düsseldorf from Leeds Bradford; Rome, Ibiza and Palma from Manchester; Murcia and Pisa from Belfast; Amsterdam, Bergen, Pisa and Menorca from Newcastle and Pisa and Murcia from Edinburgh. This resulted in a total of 31 destinations being served during the summer season. The Boeing 757s were predominantly used on the high passenger volume routes from Leeds Bradford and Manchester, offering a lower seat cost and improved operational efficiency. New routes from Blackpool to Amsterdam, Faro, Prague and Tenerife were included in the autumn schedule once the airline had three Boeing 737-300s based at the airport. Growth of the 757 fleet during the year opened the way for the introduction of new services to the Canary Islands for the winter season 2006/07, including to Tenerife from Manchester, Newcastle, Belfast and Blackpool and to Lanzarote from Leeds Bradford.

Jet2.com relocated much of its administrative and operational infrastructure from Bournemouth Airport to Leeds Bradford Airport during 2006, although the organisation and management of its operations for Royal Mail and some engineering-related administrative activities remained at Bournemouth. The company also decided to introduce in-house check-in and handling of passengers and aircraft at Leeds Bradford as well as at several airports in Spain (Malaga, Murcia and Palma) in order to provide a more efficient customer service.

In May 2005, Jet2.com added a new type to its fleet: the Boeing 757-200. The carrier's first example was 757-236 G-LSAA, which was fitted out with a high-density seating configuration for up to 235 passengers before it began commercial flights in October 2005. (Nigel Richardson)

The second Boeing 757 to join Jet2.com's fleet was 757-27B G-LSAB, which entered service in December 2005. (Nigel Richardson)

Boeing 757-23A G-LSAC was previously operated by Air Transat and Ryan International Airlines before being acquired by the Dart Group in March 2006 and entering service with Jet2.com the following May. (Nigel Richardson)

Pictured here in an unconventional Jet2.com livery is Boeing 757-236 G-LSAD. (Björn Strey, licensed under CC BY-SA 2.0)

Boeing 757-27B G-LSAE departing Manchester Airport in September 2009. This example was obtained from Air Slovakia in June 2006 and remained with Jet2.com until it was retired in December 2024. (Adrian Pingstone)

Jet2.com acquired three Boeing 757-21Bs from China Southern Airlines in November 2006, including G-LSAG, pictured here at Geneva Airport, Switzerland. (flybyeigenheer, licensed under CC BY-SA 2.0)

Boeing 757-21B G-LSAH of Jet2.com taxiing to the terminal building at Manchester Airport. (Nigel Richardson)

Acquired from China Southern Airlines in November 2006, Boeing 757-21B G-LSAI went on to operate Jet2.com's last 757 passenger service in January 2025. (Nigel Richardson)

In February 2007, the Dart Group launched Jet2holidays as a new sister company to Jet2.com. Jet2holidays is a tour operator providing holiday packages, including flights, transfers and hotels, from Jet2.com's bases which, at its inception, were Leeds Bradford, Manchester, Newcastle, Belfast, Blackpool and Edinburgh. It was anticipated that through a vertically integrated structure and by working in close association, Jet2holidays would contribute a significant proportion of the passengers on the airline's scheduled flights.

During 2007, there was a significant expansion of Jet2.com's operations: 25 new summer sun and city break routes led to a 35 per cent growth in capacity, supported by additional aircraft mainly based at Leeds Bradford and Manchester. There was an increase in services to the Canary Islands from all six bases during the winter season 2007/08 as well as additional flights to the existing skiing holiday destinations. At the beginning of the year, another 737-33A(QC), G-CELO, returned to the fleet and two aircraft were wet-leased by Jet2.com to support its scheduled operations: Boeing 737-46J, EC-KDZ, later re-registered as EI-DXO, from Futura and Boeing 737-229 Advanced, G-CEAH, from European AirCharter. Although the growth in capacity led to 3.86 million passengers being carried in 2007 compared to

2.832 million in 2006, the load factor declined from 77.2 per cent in 2006 to 73.6 per cent in 2007.

In 2008, Jet2.com, along with other airlines, faced the challenge of the UK falling into economic recession as a result of the global banking crisis and its impact on market demand. To ensure that supply was broadly equivalent to the expected demand, Jet2.com reduced its summer seat capacity by 21 per cent with some city break routes from Manchester being discontinued. Nevertheless, there was some expansion of the range of summer sun destinations with new routes introduced, including Leeds Bradford to Crete and Cyprus. For the winter season 2008/09, scheduled flight seat capacity was reduced by 49 per cent, mainly achieved by a reduction in the frequency of flights to west Mediterranean destinations. However, as part of the 2008 winter season provision, Jet2.com and Jet2holidays launched their first ever transatlantic flights to New York from Leeds Bradford Airport. Four-day winter shopping breaks were offered, with four departures in November and December 2008. The flights were operated by Jet2.com's Boeing 757s. A New York winter service was periodically offered from seven other Jet2.com bases during the subsequent ten years.

For customers, flight bookings were improved through a new reservation system, introduced in February 2008, which allowed pre-selection of seats as well as including the option of purchasing a Jet2Plus service, which provided access to airport lounges, priority check-in and a pre-ordered meal. As an additional incentive, Jet2.com launched the *myJet2* loyalty scheme in November 2008 as a means of rewarding regular customers.

The aircraft fleet was boosted by the delivery of another Boeing 757 in May 2008. An 18-year-old ex-Britannia Airways and Thomsonfly 757-236 arrived in time for the introduction of new east Mediterranean routes. It was registered G-LSAJ.

Total passenger numbers carried by Jet2.com in 2008 (3.455 million) were 11.5 per cent down on those for 2007. However, the load factor improved significantly to 79.4 per cent (2007: 73.6 per cent) reflecting the careful and effective management of available passenger capacity during the year. While passenger numbers for scheduled-only flights showed a reduction from 3.774 million in 2017 to 3.225 million in 2008, there was a 2.5 times increase in those for charter-only flights, from 86,698 in 2007 to 229,651 in 2008. An improved market demand for passenger charter services from tour operators, specialist holiday providers and for specific promotional and sporting events, together with improved use of aircraft where spare capacity existed, in particular the day-time use of the Boeing 737(QC) aircraft, contributed to the expansion of Jet2.com's charter activities. It also led to a significant contribution to the company's revenue. The Boeing 737-300(QC) aircraft was especially adaptable for charter operations, providing a range of seating capacity options and configurations to meet customer requirements, including a 78-seat, five-abreast business class arrangement through to a six-abreast, 126-seat two-class configuration in addition to the standard 149 seats of a single economy-class arrangement.

The soaring cost of jet fuel in 2008 from an average of $90 per barrel the previous year to a peak of $180 per barrel in July added to the challenges faced by airlines that year. Jet2.com hedged the cost of its fuel requirements in advance of the price rise, which minimised the impact. Nevertheless, the carrier did take action to improve fuel use through the introduction of fuel-efficiency strategies that focused on efficient aircraft loading and lower aircraft flying speeds. Another approach involved fitting blended winglets onto two Boeing 757s, which acted to reduce lift-induced drag and improved fuel efficiency by as much as five per cent. Not only were these strategies important for managing fuel use, they also reduced greenhouse gas emissions.

Jet2.com continued the expansion of its network in 2009, adding Dalaman, Dubrovnik, Tel Aviv, Split and Newquay to the list of destinations. As in the previous year, seat capacity and flight frequency were carefully managed to control potential losses, particularly in view of the decline in customer demand during off-peak periods and in the winter season. Passenger capacity was approximately seven per cent lower than in 2008. However, the number of flights operating to destinations providing both summer and winter sun holidays, such as the Canary Islands, Egypt and Cyprus, was increased. A Boeing 737-33A, G-CGET, leased from Ansett Worldwide in June 2009, was based at Manchester to supplement both scheduled and charter operations. The overall outcome for 2009 was a further 10.6 per cent decline in total passenger numbers to 3.089 million, but by balancing capacity and market demand, the load factor increased to 81.6 per cent.

The state of the economy also impacted on the passenger charter business. In 2009, Jet2.com flew 1,588 charter flight stages compared to 1,860 in 2008, carrying 195,708 passengers; an almost 15 per cent reduction with respect to the previous year. For the financial year 2009/10, the Dart Group reported a pre-tax profit of £22.2 million (2008/09: £33.5 million) and a turnover of £434 million (2008/09: £439 million), both indicative of a challenging trading environment. One positive note was the significant growth in the number of package holidays booked through Jet2holidays.com: more than 64,000 holidays were booked during the 2009/10 financial year in comparison to the relatively slow growth of the first two years for the in-house tour operator (2007/08: 34,339; 2008/09: 36,128).

Jet2.com faced a new challenge in the early part of 2010 after the eruption of the Eyjafjallajökull volcano in Iceland, which began on 14 April. The subsequent dust cloud led to the cancellation of more than 400 flights in April and May and ultimately contributed to an estimated £3 million of lost profit due to cancelled flight refunds, compensation claims and the repatriation of stranded customers.

In January 2007, Boeing 737-33A(QC) G-CELO joined the Jet2.com fleet, having previously been operated by a number of carriers including Monarch Airlines, Islandsflug and Air Atlanta Icelandic. (Pieter van Marion, licensed under CC BY-SA 2.0)

Boeing 737-46J EI-DXO was wet-leased by Jet2.com from Futura for the 2007 summer season. The aircraft initially arrived in May 2007 as EC-KDZ but was re-registered the following July. (Don Gilham)

Jet2.com leased Boeing 737-229 Advanced G-CEAH from European AirCharter between August and December 2007 to provide additional capacity. (Phil Globe)

An Ex-Britannia Airways and Thomsonfly Boeing 757-236 was acquired by Jet2.com in May 2008 and registered G-LSAJ. It is pictured here landing at Leeds Bradford Airport. (54north, licensed under CC BY-SA 3.0)

To support the expansion of its route network, Jet2.com leased Boeing 737-33A G-CGET from Ansett Worldwide between June 2009 and July 2011. (Maarten Visser, licensed under CC BY-SA 2.0)

Jet2.com acquired an additional two Boeing 737-300s in advance of the 2010 summer season. 737-3G7 G-GDFA, which had begun life operating in the USA for American West Airlines, Pace Airlines and Hooters Air, joined in January 2010. It was followed in May by an ex-Viva Air and Aegean Airlines 737-33A, registered as G-GDFB. The 757 fleet was also increased to ten aircraft when an ex-American Trans Air/ATA Airlines 757-23N was acquired in January, registered as G-LSAK. It entered service in late May on a flight from Leeds Bradford to Heraklion, Crete, following a period of maintenance.

Jet2.com's own passenger check-in and handling services were also introduced at Manchester Airport for the beginning of the summer season.

In May 2010, Jet2.com celebrated the opening of a new base at East Midlands Airport with a flight to the Cypriot resort of Paphos. One Boeing 757-200 was based at the airport for the first year, operating nine return flights each week to seven destinations, including Lanzarote and Tenerife in the Canary Islands, Dalaman, Cyprus, Corfu, Crete and Sharm El Sheikh. The opening of the East Midlands base appeared to be a fast response to easyJet's decision to suspend its flights from the airport in January 2010. Meanwhile, Jet2.com's network from the other six bases increased in 2010 to include five new destinations (Bergerac, Gran Canaria, Kos, Madeira and Monastir) and 35 new routes, of which seven were for the winter season. Hurghada in Egypt was added as a new destination for the 2010/11 winter season, while an increased number of services were flown to Geneva and Salzburg, and from Manchester to Budapest, Murcia, Prague and Rome.

A small increase in passenger seat capacity and the opening of the East Midlands base contributed to an almost 10 per cent increase in passenger numbers during 2010. Another significant factor was the further growth in the number of package holiday customers booked through Jet2holidays (just under 98,000 for the 2010/11 financial year), and hence the number of passengers on scheduled flights, together with the general increase in customer demand for seat-only bookings and seat sales to third-party tour operators. These elements also played a role in the continuing improvement of the load factor to 86.3 per cent.

The beginning of 2011 started with Jet2.com and Jet2holidays cancelling all flights and holidays to Tunisia for the summer season due to political unrest and the country being on the verge of a civil war. This was followed by a suspension of all flights and holidays to Sharm el Sheikh and Hurghada from 1 March 2011, due to civil and political unrest in Egypt. The spare capacity was reassigned to additional flights and holidays to Spain and Portugal.

Jet2.com opened a new base at Glasgow Airport at the end of March 2011. Its inaugural flight from Glasgow to Alicante on 31 March marked the introduction of services to seven destinations including Dalaman, Faro, Palma, Nice, Paphos and Tenerife along with Alicante. A ski holiday route to Geneva was added in December. In-house passenger handling was introduced at Blackpool and Newcastle airports for summer 2011 to enhance the customer experience through a more efficient service.

During the year, operations were expanded, especially from East Midlands (where new routes included Alicante, Palma, Faro, Bodrum and Rhodes), Manchester and Newcastle, with more flights to high volume, popular destinations such as the Balearic Islands, Portugal and mainland Spain. Seat capacity was increased by 29 per cent for the summer season, with several new destinations added to the summer programme including Bodrum and Krakow. For the 2011/12 winter season, capacity was increased by 15 per cent with growth provided by the first year of winter services from Glasgow and more flights to city-break destinations, such as Budapest and Prague, from Manchester. The continued suspension of flights to Egypt was compensated for by additional services to the Canary Islands for winter sun holidays.

Despite the initial problems, 2011 saw fleet modernisation and expansion taking place with the acquisition of three Boeing 737-800s. The Boeing 737-800 is a variant of the Next Generation (NG) series of the 737 family. The -800 variant first flew in July 1997 and entered service with Hapag-Lloyd Flug in April 1998. It is 19ft 11in (6.07m) longer than the 737-300, can accommodate up to 189 passengers in a single-class seating configuration and has a range of 2,935nm (5,436km). The 737-800 is also an efficient aircraft, reportedly consuming five per cent less fuel than competing aircraft at the time. Boeing 737-8K2 was leased from Volito Aviation in January 2011, having previously been operated by Transavia, and registered G-GDFC. The second aircraft, 737-8K5 G-GDFD, was acquired from TUI fly in April and the third example was an Ex-Air Europa 737-85P, delivered on lease from Fly Leasing in December and registered G-GDFF. The number of QC aircraft in the fleet increased to eight in July when Boeing 737-3Q8(QC) was acquired from TNT Express and registered G-GDFE.

Jet2.com also leased two Boeing 757-204s from Allegiant Air in March 2011, increasing the 757 fleet to 12 aircraft. The 757s, registered as G-LSAL and G-LSAM, were initially delivered as new to Britannia Airways in 1993 and had subsequently been transferred to Thomsonfly and Thomson Airways. The two aircraft featured a hybrid livery, comprising a white fuselage with 'Jet2holidays' and 'Package holidays you can trust' titles, and a blue tail, lower rear fuselage, engine nacelles and winglets. The tail also featured a bright sunburst design. Much of this paint scheme was similar to Allegiant Air's livery (owner of the 757s) and had been adopted by Jet2.com for practical reasons. However, the visual appeal of the hybrid livery and brand recognition led Jet2.com to retain and standardise this design across a number of aircraft in its fleet, even after the 757 leasing arrangement had ended.

At the beginning of the 2011 summer season, Jet2.com had a fleet of 38 aircraft, composed of 24 Boeing 737-300s (including seven QCs and one freighter), two Boeing 737-800s and 12 Boeing 757-200s. However, two leased Boeing 737-300s were returned to the lessor in July (G-CGET) and October (G-GDFA).

The positive growth trend in passenger numbers continued in 2011 with 4.236 million carried by Jet2.com, representing a 27 per cent increase on 2010 (3.339 million). It was estimated that almost eight per cent of the increase in passenger numbers during the summer season (10 per cent for the year) was due to the growth in Jet2holidays, with over 216,000 holiday packages sold during the 2011/12 financial year. Improvement in the load factor to 87.3 per cent was attributed to the combined effect of the increase in capacity and the growth in customer demand. In total, 1,031 charter flights were operated during 2011, although passenger numbers declined again to 116,277 from 175,394 in 2010, partly due to a decision to not operate charter flights during the peak months in favour of scheduled services. Nevertheless, charter flying continued to make a valuable contribution to aircraft use during the winter off-peak season and took the airline to more distant destinations than its scheduled operations, including Brazil, Canada, Florida and Sri Lanka.

Jet2.com began 2012 by being awarded 'Best Short Haul Airline' at the Travel Weekly Globe Travel Awards, voted for by travel agents and industry professionals, and overcoming competition from other carriers such as British Airways, easyJet, Iberia and Lufthansa. Jet2.com continued its objective of providing excellent customer service by introducing in-house passenger handling operations at Edinburgh and Glasgow airports in time for the 2012 winter season, while at Manchester Airport, a focus on improving operational efficiency led to the appointment of Jet2.com's own flight dispatchers.

Boeing 737-3G7 G-GDFA was acquired by Jet2.com in January 2010, having previously operated for a number of North American carriers. (Maarten Visser, licensed under CC BY-SA 2.0)

Captured departing Manchester Airport is Boeing 737-33A G-GDFB, which joined the Jet2.com fleet in May 2010. (Craig Sunter, licensed under CC BY 2.0)

Jet2.com acquired its tenth Boeing 757-200 in January 2010 when an ex- American Trans Air/ATA Airlines 757-23N arrived and was registered G-LSAK. (Pedro Aragão, licensed under CC BY-SA 3.0)

Fleet modernisation began to take place during 2011 with the acquisition of three Boeing 737-800s, a variant of the Next Generation (NG) series of the 737 family. The first example, a Boeing 737-8K2, arrived in January 2011 and was registered G-GDFC. (Nigel Richardson)

Ex-Air Europa Boeing 737-85P EC-HGP was delivered to Jet2.com in December 2011 and registered G-GDFF. (Nigel Richardson)

Jet2.com increased the number of Quick Change (QC) Boeing 737s in the fleet to eight following the acquisition of a 737-3Q8(QC) from TNT Express in July 2011, registered as G-GDFE. (Ken Fielding, https://www.flickr.com/photos/kenfielding, licensed under CC BY-SA 3.0)

Jet2.com leased two Boeing 757-204s from American low-cost carrier Allegiant Air in early 2011, including G-LSAL shown here. (John Smitherman)

Boeing 757-204 G-LSAM, on lease to Jet2.com from Allegiant Air, at Leeds Bradford Airport in July 2011. (Mark Winterbourne, licensed under CC BY 2.0)

During the 2012 summer season, Jet2.com operated services to 51 Mediterranean, Canary Islands and city break destinations from its eight bases, encompassing 173 routes. While there was a focus on high demand leisure destinations, some new destinations were added to the network, including Berlin, Istanbul and Pula (Croatia), with Grenoble (France) added as a new winter season destination. The airline also reintroduced seasonal Christmas flights to New York during the 2012/13 winter season, with flights operating from Glasgow, East Midlands and Newcastle in addition to Leeds Bradford.

From its most recent base, Glasgow Airport, Jet2.com added new services to Barcelona, Crete, Ibiza, Lanzarote, Malaga, Rhodes and Rome, supported by the addition of two more aircraft based at the airport. Five more Boeing 737s joined the fleet during the first half of the year in time for the summer operations. The first two aircraft, acquired in January and March, were Boeing 737-3Y5 G-GDFH and 737-36Q G-GDFG respectively, both of which had previously been operated by Norwegian Air Shuttle. Two more 737s delivered in March were a -300 series example, 737-36N G-GDFK, and an -800 series example, 737-804 G-GDFJ. The final example, acquired in June, was 737-36N G-GDFL, formerly of Virgin Express and Brussels Airlines. Jet2.com also took delivery of a Boeing 757-2K2 in May. The 18-year-old 757 was registered as G-LSAN but did not enter service until 8 August. Two leased 757s, G-LSAL and G-LSAM, returned to Allegiant Air at the end of the summer season.

2012 was a particularly successful year for Jet2.com. Passenger numbers grew by almost 13 per cent to 4.776 million and the load factor increased to 89 per cent, both boosted by a strong demand for holiday packages through Jet2holidays, with bookings almost doubling to 417,390 for the 2012/13 financial year (2011/12: 216,520). Overall, scheduled seat capacity was increased by 10 per cent during the year to support the anticipated growth of Jet2holidays, and Jet2.com's performance contributed to a strong financial return for the Group, with a pre-tax profit of £40.5 million for 2012/13 (2011/12: 28.1 million) and a 27 per cent increase in annual turnover to £869.2 million (2011/12: £683 million).

Jet2.com expanded its fleet to over 50 aircraft in 2013 with the acquisition of seven more Boeing 737s. Four examples were -300 series aircraft. Boeing 737-36N G-GDFM and 737-3U3 G-GDFO arrived in February (although G-GDFM didn't go into service until May), having previously operated for Brussels Airlines and Thomson Airways respectively. April marked the arrival of an ex-Air Baltic 737-33V, registered as G-GDFN, followed by 737-36Q G-GDFT in May, after operating for bmibaby for almost five years. The first two -800 series aircraft were delivered in March, registered as G-GDFP and G-GDFR. Both were 737-8Z9s variants and former Austrian Airlines aircraft. Boeing 737-85F, G-GDFV, was delivered to Manchester in May, on lease from Macquarie AirFinance. In addition, Jet2.com wet-leased a Boeing 737-8S3, OK-TSA, from Travel Service for 11 months from April 2013 and an Airbus A320-233 G-POWI from Titan Airways from May to September.

Scheduled passenger seat capacity was increased by 14 per cent over the year as the airline expanded its network to 205 routes, with much of the additional capacity assigned to popular Mediterranean and Canary Island destinations. A greater number of scheduled flights were operated during the winter season, offset by a reduction in charter services.

The Dart Group made several improvements to its infrastructure during 2013 to support the growth of its operations. In March 2013, high-quality office space in The Mint building near Leeds city centre was leased to accommodate the Jet2holidays and Jet2.com call centre and associated administrative and commercial personnel. It was aptly named 'Holiday House'. A building was also purchased in Bradford in September for the development of a flight simulator and training centre for pilots and cabin crew.

During 2013, Jet2.com operated over 36,000 stage flights carrying 5.515 million passengers at a load factor of 89.8 per cent. The year marked the end of a very successful first decade of operations for Jet2.com. Passenger numbers had grown from just under 605,000 in 2003 to over five million for the first time since its formation. The airline had expanded from a single base at Leeds Bradford Airport to eight bases, from which its network had grown from nine routes in 2003 to over 200 in 2013, supported by growth and modernisation of its fleet of aircraft. The establishment of Jet2holidays in 2007 and its subsequent expansion made a major contribution to the airline's growth, with 830,000 customers in 2013. Throughout the ten-year period, the company continued to operate nightly mail services on behalf of Royal Mail, using its fleet of Boeing 737-300(QC) aircraft. It had been operating flights for Royal Mail since 1980. In June 2013, it was announced that Jet2.com had retained contracts for six out of the eight Royal Mail routes it operated, with the revised network coming into effect in 2014.

Boeing 737-3Y5 G-GDFH of Jet2.com at Manchester Airport. (Nigel Richardson)

Boeing 737-36N G-GDFK was acquired by Jet2.com in March 2012. It had first entered commercial service with British Airways as G-XBHX in May 1998. (Anna Zvereva, licensed under CC BY-SA 2.0)

Another new addition to Jet2.com's fleet in March 2012 was Boeing 737-804 G-GDFJ. (Nigel Richardson)

An ex-Virgin Express and Brussels Airlines Boeing 737-36N was delivered to Jet2.com in June 2012, registered as G-GDFL. (Nigel Richardson)

Jet2.com took delivery of a Boeing 757-2K2 in May 2012. Registered as G-LSAN, the 18-year-old 757 did not enter into service until early August 2012. (Nigel Richardson)

Boeing 737-3U3 G-GDFO of Jet2.com at Manchester Airport in July 2014. (Nigel Richardson)

Jet2.com acquired this former British Airways, easyJet, Air Sahara and bmibaby Boeing 737-36Q in May 2013 and registered it as G-GDFT. (Nigel Richardson)

An ex-Austrian Airlines Boeing 737-8Z9 (OE-LNJ) was acquired by Jet2.com in March 2013 as G-GDFP. (Nigel Richardson)

Taxiing out to runway 23L at Manchester Airport is Jet2.com's Boeing 737-8Z9 G-GDFR. (Nigel Richardson)

Boeing 737-85F G-GDFV was delivered to Jet2.com in May 2013 on lease from Macquarie AirFinance. (Nigel Richardson)

Jet2.com wet-leased Boeing 737-8S3 OK-TSA from Travel Service in April 2013, for an 11-month period. (Alec Wilson, licensed under CC BY-SA 2.0)

Airbus A320-233 G-POWI was wet-leased by Jet2.com from Titan Airways for the 2013 summer season. (Ken Fielding, https://www.flickr.com/photos/kenfielding, licensed under CC BY-SA 3.0)

Chapter 4

Growth and Expansion (2014–23)

Jet2.com began 2014 with the acquisition of five Boeing 737-800s in time for the summer season. Two Boeing 737-8K5s, formerly of Hapag Lloyd, Air Berlin and TUI fly Germany, were registered as G-GDFW and G-GDFX and obtained on lease in March and May respectively. Two 737-86Qs delivered in April included G-GDFZ, an ex-Air Europa aircraft, and G-GDFY, which arrived from Travel Service Hungary. Both aircraft were leased from ACG. Another former Travel Service aircraft, a Boeing 737-86N, was delivered in May and registered as G-GDFS, on lease from GECAS. In October, Jet2.com took delivery of a 737-8K5 from TUI fly, which was registered as G-GDFU. The arrival of these aircraft marked the beginning of another period of fleet modernisation and replacement. In December 2014, two of Jet2.com's initial and older Boeing 737-300s, G-CELC and G-CELU, previously belonging to Channel Express, were retired.

Twelve new routes were added in 2014, together with three new destinations: Fuerteventura, Verona and Vienna. A slight dip in market demand in the early summer season, especially for eastern Mediterranean and Canary Island destinations, was concerning for Jet2.com. However, it was followed by a significant improvement in bookings for both flight-only tickets and package holidays through Jet2holidays in the latter part of the season, a trend which was maintained through the 2014/15 winter season. Passenger numbers for the calendar year exceeded 6 million (6.008 million) although there was a slight drop in the load factor to 89 per cent.

Blackpool Airport, one of Jet2.com's bases from where it operated 11 routes, was closed on 15 October 2014 due to financial problems. The two aircraft based at Blackpool and Jet2.com's services were initially transferred to Manchester Airport on 10 October.

In April 2015, Jet2holidays launched Jet2CityBreaks as a new brand, offering return flight and hotel packages to 20 European cities: Alicante, Amsterdam, Barcelona, Budapest, Cologne, Dubrovnik, Düsseldorf, Florence, Krakow, Las Palmas, Malaga, Palma, Paris, Pisa, Prague, New York, Nice, Rome, Venice and Vienna. For the 2015 summer season, new destinations, including Antalya, Kefalonia and Malta were added to Jet2.com's network, the total number of routes having increased to 227 despite the loss of the Blackpool base.

Jet2.com operated a fleet of 59 aircraft for much of 2015. Three ex-Hapag Lloyd and TUI fly Germany Boeing 737-8K5s were acquired between January and April and registered as G-JZHA, G-JZHB and G-JZHC, two of which were on lease from AerCap. A fourth aircraft, a 737-808 formerly of Air China, was delivered in April, registered as G-JZHD. To supplement the fleet for the summer season, two aircraft were wet-leased: a Boeing 757-256, EC-ISY, from Privilege Style (May–September) and an Airbus A320-232, G-POWM, from Titan Airways (April–September). At the end of the year, two more of the older 737-300s (G-CELB and G-CELD) were withdrawn from the fleet.

At the start of September 2015, Jet2.com announced that it had finalised an order with Boeing for 27 737-800s; this was the first time the carrier had bought a new aircraft directly from the manufacturer. Three more aircraft were ordered from Boeing in December 2015. The order was worth approximately $2.9 billion based on the list price, but the company had managed to negotiate a significant discount from this price. The investment aimed to support the next phase of major expansion and growth of the business through increasing and modernising the airline's fleet size to ensure enhancement of the comfort and quality available to passengers. The aircraft were due to be delivered between September 2016 and April 2018.

There was a 2.6 per cent decline in the number of passengers carried during 2015 compared to the previous year, although the load factor increased to 91.3 per cent, reflecting the cautious approach to the management of the seat availability in response to weaker customer demand at the beginning of the 2014 summer season. Capacity was reduced by one per cent overall during the 2015/16 financial year and just over 2,700 fewer stage flights were operated during the 2015 calendar year in comparison to 2014. Nevertheless, there was a 22 per cent increase in the number of Jet2holidays customers, up to 1.22 million (2014/15: 1 million) during the 2015/16 financial year, which provided 40 per cent of the transported passengers. The Dart Group's pre-tax profit increased by 82 per cent to £104.2 million (Jet2.com and Jet2holidays operating profit: £99.6 million) and revenue was £1,405.4 million (2014/15: £1,253.2 million), a 12 per cent increase on the previous financial year.

A further addition to Jet2.com's growing Boeing 737-800 fleet was 737-8K5 G-GDFW, which was acquired from TUIfly Germany in March 2014. (Nigel Richardson)

Boeing 737-86Q G-GDFZ departing from Manchester Airport in June 2015. (Nigel Richardson)

Boeing 737-86Q, previously of Travel Service Hungary, was acquired by Jet2.com as G-GDFY in April 2014. (Nigel Richardson)

Jet2.com took delivery of 737-8K5 G-GDFU from TUIfly in October 2014. (Nigel Richardson)

Three ex-Hapag Loyd and TUIfly Germany Boeing 737-8K5s were acquired by Jet2.com between January and April 2015, including G-JZHA, which was delivered in January. (Nigel Richardson)

Boeing 737-8K5 G-JZHC prepares to depart from runway 23L at Manchester Airport. (Nigel Richardson)

Jet2.com wet-leased Boeing 757-256 EC-ISY from Privilege Style during May to September 2015. (Nigel Richardson)

Airbus A320-232 G-POWM of Titan Airways was leased by Jet2.com for the 2015 summer season. (David Unsworth)

As part of Jet2.com's ongoing fleet modernisation and replacement programme, four used Boeing 737-800s were purchased in early 2016. The first two aircraft, 737-8K2s G-JZHE and G-JZHF, were obtained from Transavia in March. In May, two former Air Europa and Transaero Airlines 737-85Ps were acquired and registered as G-JZHG and G-JZHH. A Boeing 737-85P, EC-LTM, was wet-leased from Air Europa for the summer season (May–October), 737-8Q8 OY-PSA from Primera Air Scandinavia (April–October) and a Boeing 757-256 G-POWH and Airbus A321-211 G-POWN from Titan Airways (March–October).

The first of 30 brand new Boeing 737-8MG aircraft, ordered in 2015, was delivered to Jet2.com's base at Leeds Bradford Airport on 15 September 2016, registered as G-JZHJ. The delivery followed a positioning flight of almost 11 hours from Boeing Field, Seattle, USA, via Gander. The new aircraft were fitted with the 737 Boeing Sky Interior, designed to provide a more relaxing cabin environment for passengers. Originally designed for the Boeing 787 Dreamliner, Sky Interior features sculpted sidewalls and window reveals, large pivoting overhead bins that follow the curve of the cabin walls, extra legroom and LED lighting, all of which enhances the sense of spaciousness, and more headroom. Seven more 737-8MGs were delivered by the end of the year: G-JZHK and G-JZHL arrived in October, G-JZHM, G-JZHN and G-JZHO in November and G-JZHP and G-JZHR in December. Following the arrival of the eight new aircraft, Jet2.com announced in December that it had finalised an order with Boeing for four more 737-800s, bringing the total order to 34 aircraft. The four additional aircraft were due to be delivered between August 2018 and January 2019.

As the new aircraft arrived, five Boeing 737-300s were retired: G-CELS in September, G-CELZ (one of the QC examples) in October, G-CELJ and G-CELR (another QC aircraft) in November and G-CELP in December 2016.

In July 2016, it was announced that Birmingham Airport was to become Jet2.com's next base from April 2017. Two months later, London Stansted Airport was confirmed as the airline's ninth operating base, as of the beginning of April 2017.

For most of the 2016 summer season, Jet2.com operated a fleet of 62 aircraft to 63 destinations (serving 440 holiday resorts) from its seven bases. Three new destinations were added to the summer programme: Girona (Costa Brava), Naples and Berlin. Passenger numbers continued to grow in 2016, with 6.721 million passengers carried, a 14.8 per cent increase on 2015. Jet2.com reported that nearly 50 per cent of the transported passengers were Jet2holidays customers, with package holiday bookings rising to 1.73 million during the 2016/17 financial year.

In November 2016, Jet2.com opened a new maintenance hangar at Manchester Airport. The hangar, covering 6,000sq metres of floor space, allowed maintenance teams to work simultaneously on up to three Boeing 737s side-by-side or two Boeing 757-200s.

Jet2.com opened its new bases at Birmingham Airport and London Stansted Airport on 30 March 2017 with inaugural flights to Malaga and Alicante respectively. During 2017, four new Boeing 737-800s were initially based at Birmingham and operated 57 weekly flights at peak times to 23 destinations, including Alicante, Crete, Fuerteventura, Girona, Gran Canaria, Ibiza, Lanzarote, Mallorca, Malaga, Menorca, Paphos, Reus, Rhodes and Tenerife, with winter-only services for 2017/18 to Geneva, Grenoble, Krakow, Madeira, Prague, Salzburg and Turin. In November, Jet2.com operated two flights from Birmingham to New York as part of short-stay trips organised by Jet2CityBreaks.

Seven new 737-800s were based at Stansted, operating 96 weekly flights during peak periods, initially to 27 destinations, including Alicante, Antalya, Corfu, Crete, Dalaman, Faro, Fuerteventura, Girona, Gran Canaria, Ibiza, Kos, Lanzarote, Larnaca, Madeira, Mallorca, Malaga, Menorca, Paphos, Pula, Reus, Rhodes, Split, Tenerife and Zante. Winter-only services in 2017/18 included flights to Dubrovnik, Geneva and Salzburg. Jet2.com also launched four exclusive flights from London Stansted to New York in November/December, operated by an Airbus A330.

As part of Jet2.com's and Jet2holidays' largest ever summer programme, Costa de Almeria and Halkidiki were included as new destinations in 2017. Flights to Almeria were operated from Leeds Bradford, Manchester, Edinburgh, East Midlands and Newcastle, and to Halkidiki from Leeds Bradford, Manchester, East Midlands and Edinburgh. The Jet2.com fleet was increased to 75 aircraft for the summer season. Eight new Boeing 737-8MGs were delivered between 28 January and 29 April 2017 and registered as G-JZHS, G-JZHT, G-JZHU, G-JZHV, G-JZHW, G-JZHX, G-JZHY and G-JZHZ. In addition, Jet2.com acquired two used 737-800s: an ex-Air Europa 737-85P, delivered in March and entered service the following month as G-DRTA, and on 23 June, a 737-86N, previously operated by Futura, Air Sahara, JetLite and Transaero Airlines, was ferried to East Midlands and registered as G-DRTB.

In May 2017 an Airbus A330-243, G-VYGL, arrived at Manchester Airport in a hybrid Jet2 livery to begin a five-month wet-lease arrangement from AirTanker Services. This aircraft was configured in a 2-4-2 all-economy seating layout and could accommodate up to 327 passengers. It was operated by Jet2.com on the most popular routes from Manchester to Mallorca, Tenerife, Lanzarote, Ibiza and Rhodes. The aircraft was wet-leased again the following year for the summer season and from April 2019 to March 2020. Additional capacity during the 2017 summer season was also provided through the wet lease of 737-86Q EC-IDT from Air Europa between April and the end of October, and of 737-8Q8 OY-PSA from Primera Air Scandinavia from May to October. Airbus A321-211 G-POWN was wet-leased again from Titan Airways and returned under a similar arrangement in 2018, 2019, 2022, 2023 and 2024.

Boeing 737-8K2 G-JZHF was acquired by Jet2.com from Transavia in March 2016. (Nigel Richardson)

Jet2.com's Boeing 737-85P G-JZHH taxiing at Lanzarote Airport, Spain. (Rob Hodgkins, licensed under CC BY-SA 2.0)

Boeing 737-85P, EC-LTM, was wet-leased from Air Europa for the 2016 summer season (May to October). (David Montgomery)

Boeing 737-8Q8 OY-PSA was wet-leased by Jet2.com from Primera Air Scandinavia between April and October 2016. (Nigel Richardson)

Jet2.com supplemented its Boeing 757 fleet during 2016 through the lease of Boeing 757-256 G-POWH from Titan Airways. (Michael Oldfield, licensed under CC BY-SA 4.0)

A second aircraft was wet-leased from Titan Airways for the 2016 summer season in the form of Airbus A321-211 G-POWN, shown here at Manchester in September 2016. (Nigel Richardson)

Boeing 737-8MG G-JZHJ at London Stansted Airport. This aircraft was the first of an order for 34 new 737-800s placed by Jet2.com in 2015/16, and was delivered on 15 September 2016. (kitmasterbloke, licensed under CC BY 2.0)

Boeing 737-8MG G-JZHK was delivered to Jet2.com in October 2016. (Markus Eigenheer, licensed under CC BY-SA 2.0)

Jet2.com's Boeing 737-8MG G-JZHM on final approach to Geneva Airport, Switzerland. (Markus Eigenheer, licensed under CC BY-SA 2.0)

Boeing 737-8MG G-JZHN taxiing out for departure from Manchester Airport in May 2017. (Nigel Richardson)

Boeing 737-8MG G-JZHO was delivered new to Jet2.com in November 2016. (Markus Eigenheer, licensed under CC BY-SA 2.0)

Pictured on final approach to London Stansted Airport is Jet2.com's Boeing 737-8MG G-JZHR. (kitmasterbloke, licensed under CC BY 2.0)

The delivery of new Boeing 737-8MGs to Jet2.com continued during 2017 and included G-JZHU shown here taxiing at London Stansted Airport. (kitmasterbloke, licensed under CC BY 2.0)

Boeing 737-8MG G-JZHW climbing out from Innsbruck Airport, Austria, in February 2020. (BriYYZ, licensed under CC BY-SA 2.0)

Boeing 737-8MG G-JZHZ was delivered to Jet2.com in April 2017. (LRS747)

In addition to new 737s, Jet2.com acquired two used 737-800s during 2017, including an ex-Air Europa 737-85P in March, which was registered G-DRTA. (Nigel Richardson)

Boeing 737-86N G-DRTB was added to Jet2.com's fleet in June 2017, having previously been operated by Futura, Air Sahara, JetLite and Transaero Airlines. (Nigel Richardson)

Airbus A330-243 G-VYGL was wet-leased from AirTanker Services for a five-month period during the 2017 summer season and was predominantly operated by Jet2.com on its most popular routes from Manchester. (Nigel Richardson)

Boeing 737-86Q EC-IDT was wet-leased by Jet2.com from Air Europa between April and the end of October 2017. (Nigel Richardson)

Six new 737-8MGs were delivered from Boeing between October and December 2017, bringing the total received to 22 of the 34 aircraft ordered. They were registered as G-JZBA, G-JZBB, G-JZBC, G-JZBD, G-JZBE and G-JZBF. As in the previous three years, four older 737-300s were withdrawn from commercial service during 2017, including the fully-freighter configured aircraft G-CELW in January, G-CELA (a QC aircraft) in June, G-CELI in August and G-CELK in October. These withdrawals marked the end of Jet2.com's dedicated freighter operations, belying its original existence as Express Air Freight. Boeing 737-377(SF) G-CELW along with the other 737-300(QC) aircraft had been part of the fleet operating overnight flights throughout the UK for Royal Mail. The final contract, which began in 2014, expired in January 2017 whereupon Jet2.com withdrew from the tender process for a new Royal Mail contract.

A clear indication of the early success of Jet2.com's expansion plans was evident in the end of year data. The carrier operated 58,541 stage flights during 2017, a 40 per cent increase on the previous year. Passenger numbers increased by 44 per cent to 9.695 million, while the load factor remained at just over 90 per cent. Jet2.com attributed 58 per cent of the total year-on-year growth in passenger numbers within the 2017/18 financial year to the opening of the new operating bases at Birmingham and London Stansted. The financial year data also showed a 48 per cent growth in flight-only passengers, from 3.64 million in 2016/17 to 5.37 million in 2017/18.

Jet2.com took delivery of 18 more Boeing 737-800s during 2018: 11 were new 737-8MGs, of which eight examples arrived between January and the beginning of May in time for the start of the summer season and were registered G-JZBG to G-JZBN. G-JZBO, G-JZBP and G-JZBR were delivered in August, October and November respectively. The seven used 737-800s ranged between 11 and 18 years old. The first two aircraft, delivered in May and registered as G-DRTC and G-DRTD, were 737-808s, which had previously been operated by Air China and Transaero Airlines. Two more ex-Transaero Airlines aircraft, 737-85P G-DRTF and 737-8K5 G-DRTE, followed in July and a Boeing 737-86N, which had been operating in Russia for Orenair, Rossiya and Saratov Airlines for seven years, was delivered in late September (registered as G-DRTN) and initially stored for three months before entering service in December. G-DRTT, a 737-8Q8 from Primera Air, was delivered in December 2018 but didn't enter service until late February 2019, and a former Shenzhen Airlines 737-8BK was ferried to Leeds Bradford in December on lease from Avolon and registered G-DRTG. The retirement of seven Boeing 737-300s during the year included G-CELF and G-CELG in April, G-CELH in September, G-GDFH in October, and G-CELO, G-CELV and G-CELX in November.

Jet2.com operated 23 new routes in 2018/19, mainly from Birmingham and London Stansted, with overall seating capacity increased by 23 per cent to 13.81 million. To support the expansion, especially during the summer season, five aircraft were wet-leased: an Airbus A321-231 (YL-LCV) from SmartLynx Airlines from May to October, an Airbus A330-243 (G-VYGL) from AirTanker from May to November, a Boeing 757-256 (G-POWH) from May to September, two Airbus A321-211s, G-POWN from February to October and G-POWV from May to October, from Titan Airways. Almost 71,000 stage flights were operated during 2018, representing a 21.2 per cent increase on 2017. The number of passengers carried in a year exceeded 10 million for the first time. Jet2.com flew a total of 12.171 million flight-only and package holiday passengers (one-way passenger sectors) in 2018, a growth of 25.5 per cent, and the load factor improved to 91.4 per cent. Jet2holidays customers continued to grow, reaching 3.17 million during financial year 2018/19 (2017/18: 2.50 million), and were reported to represent 49 per cent of the passengers flown by the airline.

On 23 January 2019, Jet2.com took delivery of the last of the 34 new Boeing 737-800s. This latest aircraft was registered as G-JZBS. It was followed by the arrival of nine more used Boeing 737-800s throughout the year. Another ex-Shenzhen Airlines 737-8BK was acquired on lease from Avolon in January and registered G-DRTH. Two Boeing 737-86Ns were acquired in March from Primera Air Nordic via Genesis Aircraft Services. Registered as G-DRTU and G-DRTW, the latter was placed in storage in Dublin until entering service in June. An Ex-Norwegian Air International 737-8JP, registered as G-DRTO, was delivered to Manchester in late May and entered service in late June, together with a former Jet Airways 737-8FH, registered as G-DRTI, which only began commercial services with Jet2.com in mid-October. Four aircraft were delivered in the last two months of 2019: an ex-Air China Boeing 737-86N and a former Air Europa 737-85P were delivered in November, registered as G-DRTR and G-DRTM respectively, and began commercial operations with Jet2.com in January/February 2020, and two former Ryanair and Okay Airways 737-8AS variants were acquired in December, and registered as G-DRTY and G-DRTZ. G-DRTZ entered service in July 2020 but it was 12 months before G-DRTY made its debut for Jet2.com.

Boeing 737-8MG G-JZBA was delivered new to Jet2.com in October 2017. (Nigel Richardson)

Pictured at Edinburgh Airport, is Boeing 737-8MG G-JZBC (David Unsworth)

Boeing 737-8MG G-JZBD landing at Glasgow Prestwick Airport. (Robert Dowson)

Boeing 737-8MG G-JZBF taxiing to the main terminal buildings at Manchester Airport following arrival on runway 23R. (Nigel Richardson)

Boeing 737-8MG G-JZBG on final approach to Tenerife South Airport. (Alan Wilson, licensed under CC BY-SA 2.0)

Jet2.com took delivery of Boeing 737-8MG G-JZBI, shown here on final approach to Geneva Airport, Switzerland, in February 2018. (Markus Eigenheer, licensed under CC BY-SA 2.0)

Boeing 737-8MG G-JZBM was delivered as new to Jet2.com in April 2018. (Nigel Richardson)

Pictured at Corfu International Airport 'Ioannis Kapodistrias' is Boeing 737-8MG G-JZBP. (LRS747)

Boeing 737-8MG G-JZBR, delivered to Jet2.com in November 2018, is shown on final approach to London Stansted Airport. (kitmasterbloke, licensed under CC BY 2.0)

Boeing 737-808 G-DRTC was acquired by Jet2.com from Transaero Airlines in May 2018. (Gerry Barron)

Boeing 737-85P G-DRTF at Geneva Airport, Switzerland. (Markus Eigenheer, licensed under CC BY-SA 2.0)

Boeing 737-86N G-DRTN of Jet2.com departing from Faro Airport, Portugal. (Till Niermann, licensed under CC BY 3.0)

Airbus A321-231 YL-LCV was wet-leased by Jet2.com from Latvia-based SmartLynx Airlines for the 2018 summer season. (Riik@mctr, licensed under CC BY-SA 2.0)

Airbus A321-211G-POWV of Titan Airways departing from London Stansted Airport while on lease to Jet2.com for the 2018 summer season. (Alan Wilson, licensed under CC BY-SA 2.0)

In January 2019, Jet2.com took delivery of the last of 34 brand new Boeing 737-800s ordered in 2015 and 2016, taking the number of aircraft in its fleet to 100. The final aircraft to be delivered, shown here, was registered G-JZBS. (John Visanich)

Ex-Shenzhen Airlines Boeing 737-8BK was acquired by Jet2.com in January 2019 and registered G-DRTH. (LRS747)

Boeing 737-86N G-DRTU joined the Jet2.com fleet in March 2019, having previously been operated by Skymark Airlines and Primera Air Scandinavia/Nordic. (kitmasterbloke, licensed under CC BY 2.0)

Former Jet Airways Boeing 737-8FH was delivered to Jet2.com as G-DRTI in May 2019 but didn't enter service until mid-October of that year. (kitmasterbloke, licensed under CC BY 2.0)

On final approach to Birmingham Airport in July 2020 is Jet2.com's Boeing 737-85P G-DRTM. (Alec BHX/KKC, licensed under CC BY-SA 2.0)

Boeing 737-8AS G-DRTZ is awaiting its next complement of passengers at Birmingham Airport. (Alec Wilson, licensed under CC BY-SA 2.0)

Jet2.com operated a fleet of 100 aircraft for the 2019 summer season. Short-term wet-lease arrangements were used once again to provide additional capacity. They included Airbus A320-232 ES-SAP from SmartLynx Estonia (June–October), Airbus A321-211s G-POWN, G-POWU and G-POWV from Titan Airways (May–November), A321-231 YL-LCV from SmartLynx (April–November) and two Airbus A330-243s from AirTanker, with G-VYGM joining G-VYGL, which had already operated for Jet2.com during two previous peak summer seasons (2017 and 2018).

Jet2.com continued to increase its network during the 2019/20 financial year, operating an additional 26 routes and adding Chania (Crete), Izmir (Turkey) and Burgas (Bulgaria) as new destinations. Seat capacity was increased by 14 per cent for the summer season (15 per cent overall for 2019/20) with growth supported by the new routes and increased flight frequencies to popular Mediterranean, Canary Islands and European city destinations. Following a late trend of high levels of demand for winter sun holidays, Jet2.com added extra flights to the Canary Islands, Turkey, Cyprus and Greece between October and March 2020, providing an additional 170,000 seats mainly from Manchester, Birmingham, Glasgow and Newcastle. In total, 14.393 million passengers were carried by Jet2.com during 2019 (14.62 million in the financial year 2019/20); the average load factor was just below the previous year at 90.9 per cent.

The first retirement of an aircraft from the Boeing 757 fleet occurred in November 2019 when the oldest example, G-LSAG, was withdrawn from service and flown to Cotswold Airport for subsequent parting out and scrapping. Two Boeing 737-300s, G-CELE and G-GDFE, were retired in October and November.

In November 2019, Jet2.com bought peak take-off and landing slots at Manchester, Birmingham and London Stansted airports that had become available when Thomas Cook Airlines UK ceased operating on 23 September 2019 as a result of the compulsory liquidation of the Thomas Cook Group. The acquisitions provided valuable opportunities for further growth at the airports, with better flight departure times for customers. The acquisition of additional slots also allowed Jet2.com and Jet2holidays to add six new destinations to its portfolio for the 2020 summer season, namely the Greek resorts of Mykonos, Lesvos, Kalamata, Skiathos, Santorini and Preveza.

Jet2.com also made arrangements to acquire seven ex-Thomas Cook Airbus A321-211s with an average age of five years. The first one of these aircraft from Thomas Cook's German division, Condor, was delivered on 3 February 2020, on lease from Avolon, and registered G-HLYF. It was the first Airbus aircraft in Jet2.com's permanent fleet, which had previously always comprised Boeing types, although Airbus A320s and A321s had been used through short-term wet-leasing arrangements.

The A321 is the largest member of the A320 family and the first derivative of the A320, featuring a stretch of the fuselage through the insertion of a 14ft fuselage plug in front of the wing and an 8ft 9in (2.67m) plug behind the wing, thereby extending the overall length to 146ft. The original A321-100ceo (current engine option) made its maiden flight in March 1993 and entered service in January 1994 with Lufthansa. The A321-200ceo is a heavier, longer-range variant, which features more powerful engines, minor structural strengthening and two optional additional centre fuel tanks. It can carry 170–200 passengers in a standard two-class configuration and up to 218–220 passengers in a high-density economy layout, the configuration used by Jet2.com to operate over ranges of up to 3,200nm.

The outbreak of the COVID-19 pandemic in late February 2020 had a major impact on Jet2.com, along with the rest of the aviation industry. Government restrictions on travel both in the UK and across Europe led to the suspension of all flying from mid-March. Lighter travel restrictions from 10 July, following the release of an approved list of countries to which UK citizens could travel without having to self-isolate on return, allowed operations to resume. However, for Spain, the Balearic Islands and the Canary Islands, this was short-lived following the re-implementation of travel restrictions and self-isolation requirements to these regions from 26 July. Therefore, Jet2.com focused its operations on flights to destinations in the eastern Mediterranean. During the four-month period July to October 2020, Jet2.com operated 9,332 stage flights carrying 1.126 million passengers. The average load factor was 63.5 per cent, ranging from 48.1 per cent in July to 77.4 per cent in August. For the half-year period April–September 2020, Jet2.com reported 0.3 million (2019: 2.71 million) customers for Jet2holidays.

At the beginning of November 2020, travel restrictions were reintroduced in the UK for a period of four weeks following the implementation of a second national lockdown. The reintroduction of a tier system in early December led to some lifting of travel restrictions but a third national lockdown at the end of the month lasted through to March 2021, with travel abroad not permitted.

In its annual report for the financial year to 31 March 2021, Jet2 plc* reported that international travel restrictions resulted in its aircraft fleet being fully grounded for almost 29 weeks of the financial year (* The Dart Group changed its name to Jet2 plc on 16 September 2020). During the year, Jet2.com transported 1.32 million single sector passengers, a 91 per cent reduction in comparison to the previous year (2019/20: 14.62 million passengers). The decline in passenger numbers and a 90 per cent decrease in Jet2holidays customers (0.37 million in 2020/21 compared with 3.77 million in 2019/20) led to an 89 per cent reduction in revenue to £395.4 million from £3.585 billion the previous year. The company's operating loss before tax in 2020/21 was £336.1 million; in 2019/20 it had made a pre-tax profit of £293 million.

The initial response by the company to the outbreak of the pandemic was to reduce operating costs and spending and cancel or defer future additional aircraft deliveries. Cancellations included four of the ex-Thomas Cook A321-211s that Jet2.com was due to acquire on lease. During the 2020/21 financial year, Jet2 plc was able to secure an additional £1 billion in funding through the use of a credit facility from its bankers, the Covid Corporate Financing Facility provided by the Bank of England, and two equity issues which were oversubscribed by existing shareholders and new investors. The Dart Group also sold its logistics and distribution business, Fowler Welch, to the Culina Group for £98 million at the beginning of June 2020. During the year the company processed more than £1.4 billion in refunds to customers. As a result, Jet2 plc's year-end total cash balance was £1,379 million (2019/20: 1,387.5 million).

In June 2021, Jet2 plc raised an additional £537 million of funding through the issue of £387.4 million of senior unsecured convertible bonds due in 2026, together with a new unsecured £150 million loan agreement from its banks, which matured in September 2023.

Prior to the start of the COVID-19 pandemic, Jet2.com acquired another second-hand Boeing 737-800. Ex-Jet Airways 737-8AL was delivered to Manchester on 22 January 2020 and registered as G-DRTL. It remained in storage at Manchester until entering service in mid-July when travel restrictions were lifted. The onset of the pandemic led to the 'early' retirement of four aircraft from Jet2.com's fleet in March 2020: Boeing 737-300s G-CELY (the last of the QC aircraft to be withdrawn from use) and G-GDFB, and Boeing 757-200s G-LSAD and G-LSAH. The wet-lease of Airbus A330-243 G-VYGL from AirTanker was also cancelled.

Airbus A321-211s G-POWU of Titan Airways on stand at London Stansted Airport while operating for Jet2.com through a wet-lease arrangement. (kitmasterbloke, licensed under CC BY 2.0)

AirTanker's Airbus A330-243 G-VYGM joined its sister aircraft G-VYGL in operating for Jet2.com during 2019. (Nigel Richardson)

Jet2.com acquired Airbus A321-211 G-HLYF in February 2020, the first Airbus aircraft to be part of its permanent fleet. (Nigel Richardson)

Boeing 737-8AL G-DRTL was acquired by Jet2.com in January 2020, having previously been operated by Jet Airways as VT-JFJ since June 2013. (LRS747)

The continued implementation of international travel restrictions from April 2021 and the introduction of the UK government's three-tier traffic-light system for international travel in mid-May, including mandatory quarantine requirements, prevented Jet2.com from operating any scheduled flights until 24 June 2021.

At the beginning of June, Jet2 plc announced that it would restart flights and holidays to Europe from 1 July. The first flights were to the Balearic Islands (Mallorca, Menorca and Ibiza) and Madeira, all of which had been allocated to the Green Travel List, although some services had been operated to Jersey in late June. However, public confusion with the traffic-light system and a degree of nervousness around the three-weekly updates, with countries moving on and off the different travel lists (red, amber or green) depending on their perceived risk, meant that customer bookings were often delayed until close to the intended departure date.

A change in UK travel rules from 19 July allowed those travelling to amber list countries to return to England without having to quarantine, providing that passengers had been fully vaccinated with an NHS-administered vaccine in the UK at least 14 days prior to travel. In response, Jet2.com expanded its operations for the summer to 37 destinations, which led to some improvement in customer bookings. There was a return to almost 'normal' operations during October and November when the traffic-light system was replaced by a single 'red list' and measures were simplified for travel to countries on the rest of the world, or 'non-red list'. Hence, passenger numbers recovered, with 1.368 million carried in October and November, an increase of 76.6 per cent in comparison to the combined total for the usual peak period of July and August (774,853 passengers). The average load factor also improved significantly from 49.3 per cent for July and August to 76.3 per cent for October and November.

New travel measures and restrictions were implemented in the UK on 30 November 2021 in response to the emergence of the new COVID-19 Omicron variant, including the requirement to take a costly PCR test on arrival in England. Other various rules were introduced in Europe, such as Spain's ban on non-vaccinated arrivals and Portugal's requirement that anyone entering the country must have a negative COVID-19 test certificate on arrival. These travel restrictions remained in place until January 2022.

During 2021, Jet2.com operated 27,879 stage flights and carried 3.169 million passengers at an average load factor of 62.9 per cent. The company reported carrying 4.85 million passengers for the financial year 2021/22 (2020/21: 1.32 million), although these figures included February and March 2022 when public reluctance to travel lessened, leading to a gradual improvement in customer demand.

Despite the challenges presented by the COVID pandemic, Jet2.com continued with its long-term strategy to become the UK's leading leisure travel business by opening a tenth operating base at Bristol Airport on 2 July 2021, with an inaugural flight to Mallorca. When the plans to open a base at Bristol were announced in November 2020, Jet2.com intended to base three Boeing 737-800s at the airport, and operate a summer 2021 programme to 29 destinations involving up to 56 weekly flights to airports in mainland Spain, the Balearic Islands, the Canary Islands, Greece, Turkey, Italy and Portugal. It was to be followed by winter season flights to 11 destinations for winter sun and skiing holidays. Unfortunately, COVID delayed the operation of a full programme until 2022.

Jet2.com received two more Boeing 737-800s in 2021. A 737-8AL, registered as G-DRTP, was acquired on lease from AMCK Aviation on 30 June and went into service in December. The aircraft had previously been operated by Jet Airways. At the end of September, an ex-Norwegian Air and TUI fly Belgium 737-86N was delivered on lease from the FPG Group and registered as G-DRTJ. In late-November and early-December, two former Thomas Cook Airbus A321-211s were delivered on lease from ALC and registered G-HLYA and G-HLYB. These were part of a deal for seven aircraft that was arranged in November 2019 but had been subsequently modified following the impact of the COVID pandemic.

At the end of August 2021, Jet2 plc placed an initial order for 36 Airbus A321neo aircraft, with options for up to 24 more examples, as part of its fleet expansion and renewal plans. The order was a response to recent experience and evaluation of several leased A321ceo variants and represented the first time that Jet2 had placed an order for brand new Airbus aircraft.

The A321neo ('new engine option') was launched in December 2010, offering new, fuel-efficient engines, strengthened outer wing sections and landing gear, and sharklets as standard on the wing tips (instead of the previous wingtip fences) to improve lift and reduce drag contributing to a reduction in fuel burn. Jet2.com selected the CFM International LEAP-1A engine to power its A321neo fleet. The new engines can reduce fuel consumption by up to 15–20 per cent compared with the A321ceo, together with a reduction in carbon dioxide emissions and operating costs. The A321neo typically accommodates 180 to 220 passengers in a two-class configuration or up to 244 passengers in a high-density arrangement and has a range of 3,500–4,000nm (6,480–7,410km).

In October 2021, Jet2.com placed a further order for an additional 15 A321neos, taking its total firm order to 51 aircraft plus 24 options. In April 2022, it exercised purchase rights for nine of the options, resulting in a firm order for 60 examples with the potential to increase this to 75 aircraft. The deal for 60 aircraft was reported to be worth a total value of $8.1 billion based on list price, although Jet2 plc gained significant discounts. At the time of the firm order, the A321neos were due for delivery between 2023 and 2029. Three more options were converted to firm orders several months later.

From 7 January 2022, people in England who were fully vaccinated were no longer required to take a COVID test before travelling abroad. As a consequence, international travel resumed in the UK. There was a very strong demand for package holidays and Jet2holidays attracted 5.29 million customers during the 2022/23 financial year, a 40 per cent increase on pre-COVID levels (2019/20: 3.77 million). Flight-only passenger numbers were down to 5.69 million, representing a 19.4 per cent decrease on 2019/20 levels, possibly due to package holiday-only flights being operated on some popular, high volume routes. Jet2.com carried 15.533 million passengers during the 2022 calendar year at an average load factor of 87.8 per cent, and operated 92,062 stage flights. For the financial year ending 31 March 2023, Jet2 plc reported that Jet2.com transported 16.22 million passengers (2021/22: 4.85 million; 2019/2020: 14.62 million), an increase of 11 per cent in comparison to the 2019/20 financial year, just before the outbreak of the COVID pandemic. Almost 65 per cent of the passengers were Jet2holidays package-holiday customers.

Jet2.com had a fleet of 95 aircraft at the beginning of the 2022 summer season but a 14 per cent increase in seat capacity compared to 2019 required the carrier to wet-lease eight aircraft between April and October. They included two Airbus A321-211s (G-POWN and G-POWU) from Titan Airways, an A321-211 (9H-SLC) and A321-231 (9H-SLG) from SmartLynx Malta and two A321-231s provided by SmartLynx (YL-LCV and YL-LCQ). In addition, the previously-leased Airbus A330-243s (G-VYGL and G-VYGM) from AirTanker returned. An ex-flydubai Boeing 737-8KN was added to the permanent fleet as G-DRTK in March on lease from JP Lease. It entered service in May 2022.

In October 2022, Jet2 plc placed another order with Airbus for 35 A320neo aircraft, taking its combined A320neo and A321neo firm order to 98 aircraft. The order also included options for another 36 A320neo aircraft, giving Jet2.com the potential to eventually increase its combined A320neo and A321neo fleet to 146 aircraft. The new orders extended the delivery timeline up to 2031.

The Airbus A320neo was launched in 2010 and first flew in September 2014. It is powered by one of two new-generation engines, either the CFM International LEAP 1A or the Pratt & Whitney PW1100G Geared Turbofan. Both of these engines reduce fuel consumption by 15–20 per cent compared to the original A320ceo, as well as lowering carbon dioxide emissions and operating costs. The A320neo retains 95 per cent commonality with the standard A320 family. In addition to the new engines, modifications include sharklets as standard on the wing tips (instead of the previous wingtip fences), which improve lift and reduce drag, resulting in reduced fuel burn, as well as strengthened outer wing sections and landing gear, and a new 'Airspace' cabin design.

Boeing 737-8AL G-DRTP at Manchester Airport. (Nigel Richardson)

In late-September 2021, an ex-Norwegian Air and TUI fly Belgium Boeing 737-86N was delivered to Jet2.com as G-DRTJ. (Barry Swann)

Airbus A321-211 G-HLYA was the second A321 to be acquired by Jet2.com following delivery in November 2021. (Nigel Richardson)

Airbus A321-211 G-HLYB being prepared for departure at Faro Airport, Portugal. (Trevor Thornton)

Jet2.com wet-leased A321-211 9H-SLC from SmartLynx Malta for the 2022 summer season. (Manuel EstevezR)

Ex-flydubai Boeing 737-8KN A6-FEH was added to Jet2.com's fleet in March 2022 as G-DRTK. (LRS747)

Jet2.com took delivery of its first A321neo (A321-251NX) from Airbus's facility in Hamburg to East Midlands Airport on 1 April 2023, and registered it G-SUNB. The aircraft entered revenue service on 7 May, flying from Manchester Airport to Malaga. Jet2.com operates the A321neo with 232 seats, 23 per cent more than the Boeing 737-800.

For the first financial year after the pandemic (2022/23), Jet2 plc reported revenue of £5,033.5 million (2019/20: 3,584.7 million) and an operating pre-tax profit of £394 million, representing a 34.5 per cent increase on 2019/20.

Jet2.com increased its seat capacity by seven per cent on 2022 levels to provide a significantly expanded programme for the 2023 summer season. Additional flights, new routes and holidays through Jet2holidays were made available to popular destinations in the Balearic Islands, Canary Islands, Cyprus, Greece, mainland Spain, Portugal and Turkey. Much of the extra capacity was delivered by services from Birmingham, Bristol, East Midlands, Glasgow, Leeds Bradford and London Stansted airports. A total of 71,087 stage flights were operated by Jet2.com during the summer season (May–October), carrying 12.411 million (2022: 11.619 million) passengers with an average load factor of 89.5 per cent. Once again, Jet2.com added more capacity by wet-leasing ten aircraft for the busy summer season. Boeing 737-8GJ OM-MEX and 737-8BK OM-KEX were leased from AirExplore, 737-8Q8 OK-TVH and 737-82R OK-TSR from Smartwings, Airbus A321-231 CS-TRJ from Hi Fly, Airbus A330-243s G-VYGL and G-VYGM from AirTanker and Airbus A330-343 CS-WFP from World2Fly Portugal. Jet2 plc also signed a two-year, summer-season wet-lease commitment with Titan Airways for two A321-211s: G-POWN and G-POWU were allocated to Jet2.com during 2023.

There were significant additions to Jet2.com's fleet during 2023. Eleven second-hand Boeing 737-800s were acquired between February and November. Four were ex-SAS aircraft, which were delivered and registered in February–April and included a 737-883, G-JZBT, which entered service in June, and three 737-85Ps, registered as G-JZBU, G-JZBV and G-JZBW, which began commercial operations with Jet2.com in May, July and August respectively. Three different sub-variants were obtained from former Norwegian low-cost airline Flyr after it ended operations in late January 2023: two Boeing 737-82Rs, G-JZDA and G-JZDC, on lease from Pembroke, and 737-8KN G-JZDB, on lease from SDH Wings, were all delivered in March and entered service in July, and 737-8JP G-JZDD, on lease from Bank of America, was delivered in November and entered service in March 2024. Another delivery in March 2023 included a Boeing 737-8HX, previously of Sunwing Airlines, which was registered as G-JZBX and began commercial services in May, while two ex-Sun Express and Anadolujet 737-8ALs were acquired in November 2023 on lease from BOC, registered as G-DRTV and G-DRTX, and entered service in February and March 2024 respectively. Jet2.com took delivery of four more new A321-251NXs, G-SUNC in June, G-SUND in July, and G-SUNE and G-SUNF in October. The carrier also converted 12 more options for Airbus

A321-251NX aircraft into a firm commitment, increasing its total firm order to 75 examples, plus 35 A320neos. One of the Boeing 757s, 35.3-year-old G-LSAA, which the A321 was to eventually replace, was retired in November 2023 after operating in excess of 18 years for Jet2.com.

In May, Jet2 plc announced that Liverpool John Lennon Airport was to become its 11th operating base, with services scheduled to begin at the end of March 2024. Initial plans for the summer of 2024 included operating flights from Liverpool to 20 destinations in mainland Spain, the Canary Islands, the Balearic Islands, Greece, Turkey, Bulgaria, Portugal, Madeira and Cyprus, with up to 560,000 seats on sale.

To support its growth and expansion, Jet2.com enhanced its infrastructure by opening a new £8.5 million training centre for air and ground crew in Cheadle (near Manchester Airport) in February 2023, which supplemented its well-established training centre in Bradford. In October, a Retail Operations Centre was opened in Middlewich, Cheshire. It was the first of its kind for a UK airline, storing, managing and distributing in-flight retail products, including in-flight meals, gifts and duty-free items, to Jet2.com's operating bases. In addition, the company acquired additional land space at Manchester Airport in May to build a second maintenance hangar adjacent to its existing maintenance facility.

By 2023, Jet2.com had full responsibility for all ground-handling operations, including passenger check-in, baggage handling and the flight dispatcher role, at seven of its 11 UK bases, and the direct control and management of passenger check-in at ten of them.

Jet2.com's performance during the 2023 calendar year showed a 11.5 per cent increase in passenger numbers to 17.321 million (2022: 15.533 million). The carrier operated 100,817 stage flights and achieved an average load factor of 88.9 per cent.

Jet2.com took delivery of its first new A321neo (A321-251NX) on 1 April 2023, registered as G-SUNB. It entered into service the following month and is shown here operating from Manchester Airport. (Nigel Richardson)

Czech airline Smartwings supplied Jet2.com with Boeing 737-82R OK-TSR on a wet-lease arrangement for the 2023 summer season. (Roger Syratt)

Jet2.com wet-leased Airbus A321-231 CS-TRJ from Hi Fly between March and November 2023. (Don Bennett)

Airbus A330-343 CS-WFP of World2Fly Portugal at Manchester Airport in May 2023, operating on behalf of Jet2.com for the summer season. (Shaun Connor)

Boeing 737-883 G-JZBT was acquired by Jet2.com in early March 2023 and entered service just over three months later. (Stephen Brown)

Boeing 737-85P G-JZBU at East Midlands Airport, Nottingham in June 2023, shortly after beginning commercial operations with Jet2.com. (Stephen Brown)

Boeing 737-85P G-JZBV in a snowy scene at Salzburg Airport, Austria. (Richard Blata)

On final approach to Gran Canaria Airport, Spain, is Boeing 737-82R G-JZDA, which was delivered to Jet2.com in March 2023. (Fred Streep)

Boeing 737-82R G-JZDC, one of three 737-800s acquired by Jet2.com from former Norwegian low-cost airline Flyr after it ended operations in late January 2023. (Eryk Strzala)

Boeing 737-8JP G-JZDD was delivered to Jet2.com in November 2023 before entering into service in March of the following year. It had previously been operated by Norwegian Air Shuttle/International and Flyr. (Nigel Richardson)

Boeing 737-8HX G-JZBX was acquired by Jet2.com in March 2023, having previously operated for Sunwing Airlines. (Shaun Connor)

Boeing 737-8AL G-DRTV landing at Jersey Airport, Channel Islands. (Graham Hocquard)

Airbus A321-251NX G-SUNC was delivered new to Jet2.com in June 2023 and entered service before the full Jet2holidays livery had been applied. (Nigel Richardson)

Airbus A321-251NX G-SUND was acquired by Jet2.com in July 2023. (Barry Swann)

Airbus A321-251NX G-SUNE taxiing out at Manchester Airport. (Nigel Richardson)

The fifth Airbus A321-251NX to be delivered to Jet2.com was G-SUNF in October 2023. (Nigel Richardson)

Chapter 5

Jet2.com in 2024/25

In 2024, Jet2holidays attained the position of the UK's leading provider of ATOL-protected package holidays for the second year running. The tour operator was licensed to carry 6.71 million passengers travelling on package holidays, a 14.1 per cent increase on its operator capacity of 5.8 million in 2023. This achievement is central to the company's main long-term vision of being the UK's Leading and Best Leisure Travel Business.

Jet2.com remains the UK's third largest airline after British Airways and easyJet, based on the annual total of scheduled and chartered passengers. For the beginning of the 2024 summer season, the carrier had a fleet of seven Boeing 737-300s, 91 Boeing 737-800s, six Boeing 757-200s, three A321-211s and eight A321-251neos. New additions in early 2024 included two Boeing 737-8H6s (G-JZDE in January and G-JZDF in February), both ex-Pegasus Airlines aircraft obtained on lease from BOC, and a former Air Berlin and Sunwing Airlines Boeing 737-86J on lease from Avalon, registered in January as G-JZBZ. All three aircraft went into service in May 2024. Several more ex-Pegasus Boeing 737-800s joined the airline during the year: 737-8H6 G-JZDG in March, but not entering service until July, and 737-8AL G-JZDH in July, on lease from BOC, and commencing commercial services in October 2024. Three new A321-251neos, two delivered in March (G-SUNG and G-SUNH) and one in May (G-SUNJ), were available in time for the peak of the summer season, followed by a further five examples, G-SUNL, G-SUNO, G-HLYH, G-SUNM and G-HLYI in August, October, November and the latter two in December respectively.

To provide additional capacity for the summer season, Jet2.com wet-leased six Boeing 737-800s, including 737-8BKs OM-IEX, OM-KEX and OM-LEX and 737-8AS OM-JEX, from Slovakian operator AirExplore, while Croatian wet-lease specialist ETF Airways supplied 737-8K5 9A-LAB and 737-8Q8 9A-ABC from March to November. Titan Airways operated A321-211s G-POWU and G-POWN on behalf of Jet2.com, while Airbus A330-243 G-VYGM rejoined G-VYGL from AirTanker, having operated for Jet2.com throughout the 2023/24 winter season.

The growth in the fleet size (including wet-lease aircraft) from 115 in 2023 to 127 aircraft in 2024 provided a 13 per cent increase in the total available seat capacity to 14.85 million for the period 1 April to 31 September 2024. The number of passengers carried totalled 13.34 million, with an average load factor of 89.8 per cent, of which 69.2 per cent of the departing passengers were booked on package holidays, the remainder being flight-only passengers. The CAA's end of year data showed that the carrier operated 114,496 stage flights during 2024, a 13.57 per cent increase on the previous year. Annual passenger numbers increased by 12.6 per cent to 19.495 million and the average load factor was 87.9 per cent.

The increase in fleet size and seat capacity supported the addition of new routes to Jet2.com's portfolio in 2024, including Porto (from Manchester), Bergen (from Manchester, Newcastle, Leeds Bradford and Birmingham), Marrakech and Agadir in Morocco for winter sun and city breaks beginning in October 2024 (from Manchester, Birmingham, London Stansted, Glasgow, Leeds Bradford and Bristol), and Gdansk as a Christmas market destination. Jet2.com also began flying from Liverpool John Lennon Airport on 28 March 2024 with the first flights to Tenerife and Lanzarote. Four Boeing 737-800s were based at the airport for the summer period, operating up to 54 weekly flights during peak periods. A reduced 2024/25 winter season programme from Liverpool included flights to the winter sun destinations of Tenerife, Lanzarote, Gran Canaria, Fuerteventura, Alicante, Madeira, Antalya and Paphos, and to Krakow and Prague for the Christmas markets.

By early July 2024, Jet2 plc had converted its firm order for 35 A320neo plus 36 options (originally made in October 2022) to the larger A321neo variant. The following month, the company exercised its rights to firm up options for the purchase of 36 more A321-251neo, leading to a total firm commitment for 146 of the type, of which 14 have been delivered (as of April 2025). Jet2 plc originally placed an order for 36 A321neo in October 2021; this number increased to 60 aircraft by April 2022. Delivery of the A321neos will extend through to 2035. In addition, arrangements have been made for the short-term wet-lease of up to nine A321neo aircraft for 2025 and 2026. With the number of A321neo in the fleet slowly growing, Jet2.com began to retire its remaining six Boeing 757-200s at the end of the 2024 summer season. G-LSAC and G-LSAJ were both withdrawn from use on 3 November 2024 after final round trips from Manchester to Fuerteventura and Palma respectively. The aircraft were ferried to St Athan in South Wales several days later for parting out and scrapping. G-LSAN flew its last revenue service from Alicante to Manchester on 9 November before being withdrawn and shipped to St Athan. G-LSAE was retired from passenger services on 27 December 2024, followed by G-LSAK and G-LSAI on 4 and 5 January 2025 respectively. G-LSAI, the oldest passenger 757-200 in service in Europe prior to its withdrawal, operated Jet2.com's final service with the type from Geneva to Manchester on 5 January 2025.

Jet2.com was the last British airline to operate passenger-carrying Boeing 757s. It acquired its first two 757-200s in May 2005 and introduced the type into service in October 2005, operating between Leeds Bradford and Tenerife South airports. Jet2.com went on to operate 16 examples, including several through wet-lease arrangements. The versatility of the Boeing 757 played a key role in the growth and expansion of Jet2's route network. During the summer season, the type was regularly used for short-haul flights to popular, high-demand destinations in mainland Spain, the Balearic Islands and Portugal, as well as on longer routes to destinations including the Canary Islands, Greece and Turkey. In the winter season, the 757s usually operated to winter sun destinations, in particular Tenerife, Lanzarote, Gran Canaria and Fuerteventura, as well as flying skiers to Geneva. The type's arrival also allowed Jet2.com to launch winter services to New York from 2008. The Boeing 757 was a well-liked aircraft by both passengers and flight crew. However, the fleet was old, which affected its reliability, and the 757 was much less efficient than its successor, the Airbus A321neo.

In May 2024, Jet2 plc published an updated Sustainability Strategy, the first iteration having been launched in 2021. The company is committed to net zero carbon emissions by 2050 or sooner and has included a 'Climate Transition Plan' as a key component of the Sustainability Strategy. This includes a decarbonisation pathway, with the target of achieving a 35 per cent reduction in carbon emissions per revenue-paying passenger kilometre (gCO_2/RPK) by 2035 compared to 2019 (from 67gCO_2/RPK in 2019 to 43.55gCO_2/RPK in 2035). The actions that will enable Jet2.com to achieve this target include:

- An increase in the number of Airbus A321-251neo aircraft in the fleet. The company has placed a firm order for 146 examples and should have at least 100 in service by 2030. The A321neo is approximately 20 per cent more efficient at reducing fuel consumption and carbon dioxide emissions than the older aircraft in the fleet.
- The withdrawal from service of the remaining seven Boeing 737-300s and a gradual reduction in the number of older Boeing 737-800s in the fleet over the next 5–6 years, both types being less efficient than the A321neo.
- Retrofitting up to 50 Boeing 737-800s with more aerodynamic split scimitar winglets, which are expected to reduce fuel consumption by up to 1.8 per cent.
- Increasing the use of Sustainable Aviation Fuel (SAF) in the total fuel mix to a minimum of 15 per cent by 2035. During 2024, Jet2.com operated a number of flights from London Stansted, Bristol and Malaga airports using a fuel mix containing one per cent SAF. A UK government sustainable aviation fuel mandate (the UK SAF Mandate) published in April 2024 requires two per cent of the jet fuel supplied in the UK in 2025 to be SAF, rising to 10 per cent SAF by 2030. Similarly, the EU's ReFuelEU mandate requires fuel suppliers to supply, and airlines to uplift, increasing proportions of SAF for use on flights departing from EU airports. For Jet2.com to achieve this objective, it will be dependent of the production and wider availability of SAF and will require industry and governmental support.
- Increased electrification of Jet2.com-owned ground-support equipment, leading to a 99 per cent reduction in emissions by 2035. In its Annual Report 2024, Jet2 plc stated that more than 50 per cent is currently electrified.
- Further development of Free Route Airspace and Systemisation in the UK as part of the Civil Aviation Authority (CAA) Airspace Modernisation Strategy,

and similar improvements in Europe through the EU's Single European Sky initiative, both of which should lead to more direct routes and reduce flight times, costs, fuel burn and carbon dioxide emissions.

In March 2024, Jet2 plc announced that Bournemouth Airport was to become its twelfth operating base. Flights were planned to start in early April 2025, but due to high customer demand, the launch was brought forward to February to include winter sun flights to Gran Canaria, Lanzarote and Tenerife. The inaugural flight from Bournemouth to Tenerife departed on 4 February 2025. Over 270,000 tickets are available for a 2025 summer programme from Bournemouth Airport, where a fleet of two aircraft will operate up to 27 weekly flights to 16 destinations in mainland Spain, the Balearic Islands, Portugal, the Canary Islands, Madeira, Greece and Turkey.

The Bournemouth announcement was followed up in November 2024 by news of further expansion of Jet2.com and Jet2holidays with the launch of operations from London Luton Airport in April 2025. The development serves the large customer catchment area of London, Bedfordshire, Buckinghamshire and Hertfordshire, providing a summer programme that includes 17 destinations in mainland Spain, the Balearic Islands, the Canary Islands, Portugal, Italy, Greece and Turkey. Two new Airbus A321neos were initially based at the airport, operating up to 36 flights per week. The inaugural flight departed to Mallorca on 1 April 2025, followed by several more flights on the same day to Tenerife and Antalya.

In anticipation of the new bases opening at Bournemouth and Luton airports, Jet2holidays increased its ATOL licence in October 2024 to just over 7 million passengers annually travelling on package holidays. Jet2.com also increased its seat capacity for the 2025 summer season by 8.5 per cent to 18.65 million, with the two new bases contributing over 700,000 (3.75 per cent) of these seats. Expansion of its route network included the introduction of flights from Birmingham and Manchester to Salerno Costa d'Amalfi Airport in Italy for new holiday destinations along the Amalfi Coast, five weekly services from Birmingham, Manchester and London Stansted to Pula Airport in Croatia for Pula and the Istrian Coast, and weekly flights to Jerez Airport in Spain for the Costa de La Luz region from Birmingham, Leeds Bradford and Manchester airports; Jerez became the 14th Spanish airport to be used by Jet2.com.

Jet2.com had a permanent fleet of 117 aircraft at the beginning of April 2025, comprising three Airbus A321-211s, 14 Airbus A321-251NX, seven Boeing 737-300s and 93 Boeing 737-800s. The only new addition, in early 2025, was Airbus A321-251NX G-HLYJ, delivered in February. Also in February 2025, Jet2 plc announced that it expected to increase its A321neo fleet to 23 examples by the end of the summer season through the acquisition of new owned and leased Airbus A321-251NX aircraft. However, the delivery of new aircraft continued to be delayed due to supply chain problems at Airbus, requiring Jet2.com to wet-lease additional capacity to cover the gaps in available aircraft for the peak summer flying programme, leading to additional operating costs. The wet-leased aircraft that arrived in March/April included Boeing 737-8K5 9A-LAB and 737-8Q8 9A-ABC of ETF Airways, 737-8BK OM-LEX, 737-82R OM-OEX and 737-8AS OM-JEX of AirExplore, Titan Airways' A321-211s G-POWU and G-POWN and Airbus A330-243 G-VYGM from AirTanker. Airbus A330-243 G-VYGL remained on lease to Jet2.com from AirTanker and a number of rotations during the summer were also operated by another of AirTanker's A330-243s, G-VYJK. Hi Fly Malta also leased A330-202 9H-MFS to Jet2.com from late June to early September 2025. All the wet-leased aircraft were returned to their respective lessors at the beginning of November 2025.

By the end of 2025, Jet2.com's A321-251NX fleet had grown to 24 examples following the delivery of ten aircraft (G-HLYL, G-HLYK, G-SUNP, G-SUNS, G-HLYM, G-SUNU, G-SUNR, G-SUNV, G-SUNW and G-HLYN) between May and November.

On 12 November 2025, Jet2.com announced that London Gatwick Airport was to become its 14th UK base. Flights to 29 holiday destinations in mainland Spain, the Balearic Islands, the Canary Islands, Portugal, Greece, Turkey, Malta, Cyprus, Italy, Croatia and Bulgaria are due to begin on 26 March 2026. A fleet of six aircraft will operate the services, of which five A321-251NXs will be based at Gatwick Airport.

As Jet2.com progresses through its 23rd year as a passenger-carrying airline, its clear focus on customer satisfaction, the provision of package holidays, strategic expansion, fleet modernisation and environmental sustainability place it in a strong position for further growth and success in the future.

Boeing 737-8H6 G-JZDF was one of two ex-Pegasus Airlines aircraft acquired by Jet2.com early in 2024. (Shaun Connor)

On final approach to Manchester Airport in June 2024 is Boeing 737-86J G-JZBZ, which joined the Jet2.com fleet in January 2024. (Shaun Connor)

Boeing 737-8H6 G-JZDG at Glasgow International Airport in September 2024. This aircraft was acquired by Jet2.com in March 2024 but didn't enter service until July that year. (John Murdoch)

Boeing 737-8AL G-JZDH pictured landing at Bristol Airport in November 2024. (Kev Slade)

Airbus A321-251NX G-SUNG was delivered to Jet2.com in March 2024 and entered service in May with an incomplete paint scheme. (Anton Homma)

Airbus A321-251NX G-SUNH on final approach to Gran Canaria Airport, Canary Islands, in July 2024. (Roman Valladares)

Airbus A321-251NX G-SUNJ was acquired by Jet2.com in May 2024 in time for the summer season. (Richard Jackson)

To provide additional capacity for the 2024 summer season, Jet2.com wet-leased a number of aircraft including Boeing 737-8K5 9A-LAB from ETF Airways. (Barry Swann)

Another aircraft wet-leased for the 2024 summer season was Boeing 737-8Q8 9A-ABC from Croatian lease specialist ETF Airways, pictured here at Manchester Airport. (Nigel Richardson)

Jet2.com wet-leased four Boeing 737-800s from Slovakian carrier AirExplore during the summer of 2024, including 737-8BK OM-IEX. (Richard Jackson)

Boeing 737-8AS OM-JEX of AirExplore operating for Jet2.com during the summer 2024 through a wet-lease arrangement. (Richard Jackson)

Airbus A321-251NX G-SUNL on departure from Manchester Airport. (Shaun Connor)

Airbus A321-251NX G-SUNO landing at Glasgow Prestwick Airport, in January 2025. (Ian Howat)

Airbus A321-251NX G-HLYH was delivered to Jet2.com at the end of November 2024 and is shown here on final approach to Verona Villafranca Airport, Italy. (Varani Ennio)

Airbus A321-251NX G-SUNM, shown here at Birmingham Airport, was delivered new to Jet2.com in December 2024. (Richard Jackson)

Airbus A321-251NX G-HLYI, the eighth new A321neo to be delivered to Jet2.com during 2024. (Javier Rodriguez)

Boeing 757-21B G-LSAI at Geneva Airport, Switzerland, prior to its final flight back to Manchester and the last passenger flight by a Jet2.com Boeing 757. (Thomas Haynes/ Key Publishing)

Airbus A321-251NX G-HLYJ on the ground at Manchester Airport, following delivery to Jet2.com in February 2025. (Vin LK)

Airbus A321-251NX G-HLYL was delivered to Jet2.com on 22 May 2025. (Shaun Connor)

Jet2.com took delivery of its 20th A321-251NX G-SUNU, at the end of July 2025. It is shown here landing at Manchester Airport the following month. (Shaun Connor)

Fleet Lists

Abbreviations used: C/n=construction number; Lsd to = leased to; Lsd from = Leased from; rtl = return to lessor; wfu = withdrawn from use; pwfu = permanently withdrawn from use; Opf = operated for; Opby = operated by; Rtl = returned to lessor

Sources

airfleets.net
CAA G-INFO website
planelogger.com
planespotters.net

Express Air Freight/Express Air Services (EAF/EAS)

Aircraft type	C/n	Registration	Date	Notes/Fate
De Havilland Canada DHC-6-310 Twin Otter	578	G-RBLA	Jun 1982–Jan 1983	To Channel Express
Douglas DC-3 Dakota	9043	G-AKNB	Jan 1977–Oct 1978	Opby Intra Airways; to Clyden Airways
	32872	G-AMPZ	Jan 1977–Sep 1978	Opby Intra Airways; to Clyden Airways
	13468	G-AMHJ	Jan 1980–Feb 1982	To Air Atlantique
	26569	G-AMPY	Jan 1980–Feb 1982	To Air Atlantique
Handley Page HPR-7 Dart Herald Series 200	189	G-ATDS	Aug 1977–Nov 1984	To Channel Express
	174	G-BEZB	Aug 1977–Nov 1984	To Channel Express
	195	G-BFRJ	Apr 1978–Mar 1983	To Channel Express
	197	G-BFRK	Apr 1978–Jan 1983	To Channel Express
Vickers Viscount - 800	341	G-BAPE	Mar–Oct 1980	To Southern International
	344	G-BAPG	Aug 1980–Jan 1981	Lsd to Southern International
	375	G-AVJB	Jan 1980	To Jersey European

Channel Express

Type	C/n	Registration	Date	Notes/Fate
Airbus A300-B4-103(F)	124	G-CEXC	Jul 1997–Feb 2000	To TNT Express as G-TNTS (opby Channel Express until Oct 2000)
	105	G-CEXK	Apr 2002–Jan 2006	wfu Oct 2005; rtl
Airbus A300-B4-203(F)	117	G-CEXH	Apr 1998–Nov 2005	Acquired by Dart Group in July 1997 as D-ASAZ; to Kuzu Airlines Cargo
	121	G-CEXI	Sep 1998–Feb 2006	Acquired by Dart Group in Feb 1998 as D-ASAA; to TNT Express (opby Channel Express until Jun 2006)
	155	G-TNTI	Oct 1999	To TNT Express (opby Channel Express until Jul 2000)
	147	G-CEXJ	Mar 2000–Mar 2006	Wfu Oct 2005; rtl
		Boeing 737-300		
Boeing 737-330(QC)	23522	TF-ELP/ G-CELP	May 2001–Jan 2006	Lsd from Islandsflug as TF-ELP (May 2001–Oct 2003); to Jet2.com
	23523	TF-ELR/ G-CELR	May 2001–Jan 2006	Lsd from Islandsflug (May 2001–Nov 2003); lsd to Flyglobespan (Mar 2003–Jan 2006); to Jet2.com
Boeing 737-33A(QC)	25401	TF-ELA	Apr 2005	Lsd from Air Atlanta Icelandic; to Jet2.com (opby Channel Express until Jan 2006)
Boeing 737-377	23660	G-CELS	May–Dec 2002	To Jet2.com (opby Channel Express until Jan 2006)
	23657	G-CELU	Jun 2002–Apr 2003	To Jet2.com (opby Channel Express until Jan 2006)
	23661	G-CELV	Oct 2002–Feb 2003	To Jet2.com (opby Channel Express until Jan 2006)

Type	C/n	Registration	Date	Notes/Fate
Boeing 737-377(QC)	23663	G-CELA	Aug 2003–Feb 2005	To Jet2.com (opby Channel Express until Jan 2006)
	23654	G-CELX	Mar–Apr 2003	To Jet2.com (opby Channel Express until Jan 2006)
	23662	G-CELY	Apr 2004	Acquired by Dart Group in Apr 2003; to Jet2.com (opby Channel Express until Jan 2006)
	23658	G-CELZ	Sep 2003–Aug 2004	To Jet2.com (opby Channel Express until Jan 2006)
Boeing 737-377(SF)	23659	G-CELW	Oct 2003–Jan 2006	Acquired by Dart Group in Aug 2002 as N659DG; to Jet2.com
Boeing 757-200				
Boeing 757-236	24122	G-LSAA	May–Oct 2005	To Jet2.com (opby Channel Express until Jan 2006)
Boeing 757-27B	24136	G-LSAB	May–Dec 2005	To Jet2.com (opby Channel Express until Jan 2006)
De Havilland Canada DHC-6-310 Twin Otter				
De Havilland Canada DHC-6-310 Twin Otter	578	G-RBLA	Jan–May 1983	To Nordic Oil Services
Fokker F27 Friendship				
Fokker F27-500 Friendship	10503	G-CEXA	Jan 1996–Oct 2004	Pwfu
	10550	G-CEXB	Nov 1995–Mar 2005	
	10654	G-CEXE	Apr 1997–Jun 2005	
	10660	G-CEXF	Apr 1997–Oct 2004	
	10459	G-JEAP/ G-CEXG	Oct 2000–Mar 2005	
Fokker F27-600 Friendship	10349	G-FEAE	Feb 1990–May 1992	
	10405	G-BNIZ	Nov 1994–Feb 2005	
	10351	G-CEXD	Feb 1997–Oct 2004	Pwfu
	10508	G-CHNL	Mar 1995–Jan 1999	Destroyed

Type	C/n	Registration	Date	Notes/Fate
Handley Page HPR-7 Dart Herald Series 200				
Handley Page HPR-7 Dart Herald 209	189	G-ATDS	Jan 1983–Jan 1991	Wfu
	174	G-BEZB	Jan 1983–Nov 1987	Wfu
	195	G-BFRJ/ G-CEXP	Jan 1983–Mar 1984; Oct 1987–Mar 1996	Converted to freighter and re-registered G-CEXP, Oct 1987; wfu
	197	G-BFRK/ G-GNSY	Jan 1983–Oct 1984; Jun 1987–Dec 1996	Converted to freighter and re-registered G-GNSY, Jun 1987; wfu
Handley Page HPR-7 Dart Herald 210(F)	173	G-SCTT	Aug 1988–Feb 1997	Wfu
	188	G-STVN	Aug 1988–Apr 1997	Wfu
Handley Page HPR-7 Dart Herald 213	176	G-AVPN	Jan 1992–Jul 1997	Wfu
	179	G-AYMG	May 1989–Jul 1992	Wfu
Handley Page HPR-7 Dart Herald 214	185	G-ASVO	Nov 1993–Apr 1997	Damaged beyond repair Apr 1997
	186	G-BEBB/ G-CEAS	Apr 1985–Feb 1997	Re-registered as G-CEAS in Jan 1986; Converted to freighter in Dec 1987; wfu
Handley Page HPR-7 Dart Herald 214(F)	194	G-DGLD	Oct 1991–Dec 1992	Wfu
Handley Page HPR-7 Dart Herald Series 400				
Handley Page HPR-7 Dart Herald 401	187	G-BEYK	Jun–Dec 1988	Lsd from Nordic Oil Services; rtl
Handley Page HPR-7 Dart Herald 401(F)	175	G-BEYF	Aug 1988–Apr 1999	Wfu

Type	C/n	Registration	Date	Notes/Fate
Lockheed Electra				
Lockheed L-188A(F) Electra	1068	N5535/ G-CHNX	Oct 1990–Oct 1993; Nov 1994–Oct 2001	Re-registered G-CHNX in Nov 1994; wfu
	1104	N284F	Jan–Dec 1994	Lsd from Zantop International Airlines; rtl
	1107	N285F	Sep 1998–Mar 1999	Lsd from Renown Aviation; damaged beyond repair Mar 1999
	1035	N341HA	Sep 1991–May 1992; Aug 1992–Jan 1994; Jul 1995–Apr 1997	Lsd from Zantop International Airlines; rtl
	1053	N343HA	Nov 1990–Apr 1991; May 1992–Jun 1993; Mar 1994–Apr 1997	Lsd from Zantop International Airlines; rtl
	1038	N344HA	Jun–Nov 1993; Jan 1994–Jul 1995	Lsd from Zantop International Airlines; rtl
Lockheed L-188C(F) Electra	2006	G-BYEF	Dec 1998	Purchased for spare parts only, never operated
	1091	N5539/ G-CEXS	Oct 1990–May 2003	Re-registered G-CEXS in Apr 1992; to Air Spray
	2014	G-FIZU	Mar 1999–Oct 2001	Lsd from Air Atlantique/Atlantic Airlines; rtl
	1075	N347HA/ G-OFRT	Nov 1989–Apr 2003	Re-registered G-OFRT in Oct 1991; wfu
	1112	N360Q	Sep–Nov 1993; Feb 1997–Dec 1998; Mar–Dec 1999	Lsd from JBQ Aviation and Renown Aviation (1999); rtl

Jet2.com

Type	C/n	Registration	Date	Notes/Fate
Airbus A320-200				
Airbus A320-232	2564	G-POWM	Apr–Sep 2015	Lsd from Titan Airways; rtl
	1183	ES-SAP	Jun–Oct 2019	Lsd from SmartLynx Estonia; rtl
Airbus A320-233	2791	G-POWI	May–Sep 2013	Lsd from Titan Airways; rtl
Airbus A321-200				
Airbus A321-211	4461	9H-SLC	Apr–Oct 2022	Lsd from SmartLynx Malta
	3830	G-POWN	Mar–Sep 2016; Apr–Oct 2017; Feb–Oct 2018; May–Nov 2019; Apr–Oct 2022; Mar–Nov 2023; Apr–Nov 2024; Apr–Nov 2025	Lsd from Titan Airways
	3708	G-POWU	Apr–Nov 2019; Apr–Oct 2022; Apr–Nov 2023; Mar–Nov 2024; Apr–Nov 2025	Lsd from Titan Airways
	3749	G-POWV	May–Oct 2018; May–Nov 2019	Lsd from Titan Airways; rtl
	6038	G-HLYA	Nov 2021–present	Lsd from ALC
	6056	G-HLYB	Dec 2021–present	Lsd from ALC
	6639	G-HLYF	Feb 2020–present	Lsd from Avolon
Airbus A321-231	1946	9H-SLG	Apr–Oct 2022	Lsd from SmartLynx Malta; rtl
	1004	CS-TRJ	Mar–Nov 2023	Lsd from Hi Fly; rtl
	2211	YL-LCQ	Apr–Oct 2022	Lsd from SmartLynx; rtl
	2216	YL-LCV	May–Oct 2018; Apr–Nov 2019; Apr–Oct 2022	Lsd from SmartLynx; rtl

Type	C/n	Registration	Date	Notes/Fate
Airbus A321-251NX	11379	G-SUNB	Apr 2023–present	Entered service May 2023
	11489	G-SUNC	Jun 2023–present	Entered service Jul 2023
	11564	G-SUND	Jul 2023–present	
	11638	G-SUNE	Oct 2023–present	Entered service Nov 2023
	11720	G-SUNF	Oct 2023–present	
	11799	G-SUNG	Mar 2024–present	Entered service May 2024
	11806	G-SUNH	Mar 2024–present	
	11943	G-SUNJ	May 2024–present	
	12126	G-SUNL	Aug 2024–present	
	12060	G-SUNO	Oct 2024–present	
	12295	G-HLYH	Nov 2024–present	
	12127	G-SUNM	Dec 2024–present	
	12161	G-HLYI	Dec 2024–present	
	12355	G-HLYJ	Feb 2025–present	
	12553	G-HLYL	May 2025–present	
	12529	G-HLYK	June 2025–present	
	12599	G-SUNP	July 2025–present	
	12646	G-SUNS	July 2025–present	
	12667	G-HLYM	Jul 2025–present	
	12720	G-SUNU	Jul 2025–present	
	12641	G-SUNR	Sep 2025–present	
	12730	G-SUNV	Sep 2025–present	
	12760	G-SUNW	Sep 2025–present	
	12856	G-HLYN	Nov 2025– present	
Airbus A330-202	700	9H-MFS	Jun–Sep 2025	Lsd from Hi Fly Malta
Airbus A330-243	1555	G-VYGL	May–Oct 2017; May–Nov 2018; Apr 2019–Mar 2020; Apr–Oct 2022; Apr 2023–Nov 2025	Lsd from AirTanker
	1601	G-VYGM	May–Nov 2019; May–Oct 2022; Mar–Dec 2023; Apr–Nov 2024; Apr–Nov 2025	Lsd from AirTanker

Type	C/n	Registration	Date	Notes/Fate
Boeing 737-200 Advanced				
Airbus A330-343	1097	CS-WFP	Apr–Oct 2023	Lsd from World2Fly Portugal; rtl
Boeing 737-229 Advanced	21135	G-CEAH	Aug–Dec 2007	Lsd from European Aircharter; rtl
Boeing 737-300				
Boeing 737-330	23529	G-CELJ	Dec 2004–Nov 2016	Opby Channel Express until Jan 2006; pwfu
	23530	G-CELK	Feb 2005–Oct 2017	Opby Channel Express until Jan 2006; pwfu
	23525	G-CELH	Aug 2004–Sep 2018	Opby Channel Express until Jan 2006; pwfu
	23526	G-CELI	Sep 2004–Aug 2017	Opby Channel Express until Jan 2006; pwfu
Boeing 737-330(QC)	23522	G-CELP	Jan 2006–Dec 2016	Pwfu
	23523	G-CELR	Jan 2006–Nov 2016	Lsd to Flyglobespan Jan 2006–May 2007; pwfu
Boeing 737-33A	23831	G-CELC	Jul 2003–Dec 2014	Opby Channel Express until Jan 2006; pwfu
	23832	G-CELD	Jul 2003–Dec 2015	Opby Channel Express until Jan 2006; pwfu
	24029	G-CELE	Jul 2003–Oct 2019	Opby Channel Express until Jan 2006; pwfu
	27455	G-CGET	Jun 2009–Jul 2011	Lsd from Ansett Worldwide (AWAS); rtl
	25743	G-GDFB	May 2010–Mar 2020	Lsd from AWAS until Dec 2011; pwfu
Boeing 737-33A(QC)	24028	G-CELO	Feb–Jun 2006; Jan 2007–Nov 2018	Pwfu
	25401	TF-ELA	Apr 2005–Jun 2006	Lsd from Air Atlanta Icelandic; opby Channel Express until Jan 2006; rtl
Boeing 737-33V	29332	G-GDFN	Apr 2013–present	
Boeing 737-36N	28572	G-GDFK	Mar 2012–present	
	28568	G-GDFL	Jun 2012–present	
	28586	G-GDFM	Feb 2013–present	Entered service May 2013

Type	C/n	Registration	Date	Notes/Fate
Boeing 737-36Q	28658	G-GDFG	Mar 2012–present	
	29141	G-GDFT	May 2013–present	
Boeing 737-377	23664	G-CELB	Jul 2003–Nov 2015	Opby Channel Express until Jan 2006; pwfu
	24302	G-CELF	Oct 2004–Apr 2018	Opby Channel Express until Jan 2006; pwfu
	24303	G-CELG	Oct 2004–Apr 2018	Opby Channel Express until Jan 2006; pwfu
	23660	G-CELS	Dec 2002–Sep 2016	Opby Channel Express until Jan 2006; pwfu
	23657	G-CELU	Apr 2003–Dec 2014	Opby Channel Express until Jan 2006; pwfu
	23661	G-CELV	Feb 2003–Nov 2018	Opby Channel Express until Jan 2006; pwfu
Boeing 737-377(QC)	23663	G-CELA	Feb 2005–Jun 2017	Opby Channel Express until Jan 2006; pwfu
	23654	G-CELX	Apr 2003–Nov 2018	Opby Channel Express until Jan 2006; pwfu
	23662	G-CELY	Apr 2004–Mar 2020	Opby Channel Express until Jan 2006; pwfu
	23658	G-CELZ	Aug 2004–Oct 2016	Opby Channel Express until Jan 2006; pwfu
Boeing 737-377(SF)	23659	G-CELW	Jan 2006–Jan 2017	Pwfu
Boeing 737-3G7	24011	G-GDFA	Jan 2010–Oct 2011	Pwfu
Boeing 737-3Q8(QC)	24131	G-GDFE	Jul 2011–Nov 2019	Pwfu
Boeing 737-3U3	28740	G-GDFO	Feb 2013–present	
Boeing 737-3Y5	25615	G-GDFH	Jan 2012–Oct 2018	Pwfu
Boeing 737-400				
Boeing 737-46J	27826	EC-KDZ/EI-DXO	May–Oct 2007	Lsd from Futura; rtl
Boeing 737-4Y0	23980	TF-ELZ	Feb 2005–Jan 2006	Lsd from Islandsflug; rtl

Type	C/n	Registration	Date	Notes/Fate
Boeing 737-800				
Boeing 737-804	28229	G-GDFJ	Mar 2012–present	Lsd from ACG
Boeing 737-808	34702	G-DRTC	May 2018–present	
	34703	G-DRTD	May 2018–present	
	34706	G-JZHD	Apr 2015–present	
Boeing 737-82R	40014	G-JZDA	Mar 2023–present	Lsd from Pembroke until Nov 2023, then AviLease; entered service July 2023
	40881	G-JZDC	Mar 2023–present	Lsd from Pembroke until Nov 2023, then AviLease; entered service July 2023
	40875	OK-TSR	Apr–Nov 2023	Lsd from Smartwings; rtl
	38173	OM-OEX	Apr–Nov 2025	Lsd from AirExplore
Boeing 737-85F	28821	G-GDFV	May 2013–present	Lsd from Macquarie AirFinance
Boeing 737-85P	28385	G-GDFF	Dec 2011–present	Lsd from Fly Leasing until Jun 2015, then BBAM
	36591	EC-LTM	May–Oct 2016	Lsd from Air Europa; rtl
	33972	G-DRTA	Mar 2017–present	Entered service Apr 2017
	28387	G-DRTF	Jul 2018–present	
	33981	G-DRTM	Nov 2019–present	Entered service Feb 2020
	35706	G-JZBU	Feb 2023–present	Entered service May 2023
	35707	G-JZBV	Mar 2023–present	Entered service Jul 2023
	35708	G-JZBW	Apr 2023–present	Entered service Aug 2023
	28388	G-JZHG	May 2016–present	
	28536	G-JZHH	May 2016–present	
Boeing 737-86J	37752	G-JZBZ	Jan 2024–present	Lsd from Avolon
Boeing 737-86N	28610	G-DRTB	Jun 2017–present	
	35647	G-DRTJ	Sep 2021–present	Lsd from FPG Group
	32735	G-DRTN	Sep 2018–present	Entered service Dec 2018
	35209	G-DRTR	Nov 2019–present	Entered service Jan 2020
	34247	G-DRTU	Mar 2019–present	
	28618	G-DRTW	Mar 2019–present	Entered service Jun 2019
	32243	G-GDFS	May 2014–present	Lsd from GECAS until Nov 2021, then AerCap

Type	C/n	Registration	Date	Notes/Fate
Boeing 737-86Q	32773	EC-IDA	May–Oct 2018	Lsd from Air Europa; rtl
	30281	EC-IDT	Apr–Oct 2017	Lsd from Air Europa; rtl
	30278	G-GDFY	Apr 2014–present	Lsd from ACG
	30276	G-GDFZ	Apr 2014–present	Lsd from ACG; entered service May 2014
Boeing 737-883	34546	G-JZBT	Mar 2023–present	Entered service Jun 2023
Boeing 737-8AL	39059	G-DRTL	Jan 2020–present	Lsd from MCAP; entered service Jul 2020
	39067	G-DRTP	Jun 2021–present	Lsd from AMCK Aviation; entered service Dec 2021
	40554	G-DRTV	Nov 2023–present	Lsd from BOC; entered service Feb 2024
	40555	G-DRTX	Nov 2023–present	Lsd from BOC; entered service Mar 2024
	62687	G-JZDH	Jul 2024–present	Lsd from BOC; entered service Oct 2024
Boeing 737-8AS	33554	G-DRTY	Dec 2019–present	Entered service Dec 2020
	33557	G-DRTZ	Dec 2019–present	Entered service Jul 2020
	29932	OM-JEX	Mar–Nov 2023; Mar–Nov 2024; Apr–Nov 2025	Lsd from AirExplore
Boeing 737-8BK	33020	G-DRTG	Dec 2018–present	Lsd from Avolon
	33828	G-DRTH	Jan 2019–present	Lsd from Avolon
	33016	OM-IEX	Mar–Nov 2024	Lsd from AirExplore; rtl
	33018	OM-KEX	Apr–Oct 2023; Apr–Nov 2024	Lsd from AirExplore; rtl
	29642	OM-LEX	Mar–Nov 2024; Mar–Nov 2025	Lsd from AirExplore
Boeing 737-8FH	29639	G-DRTI	May 2019–present	Entered service Oct 2019
Boeing 737-8GJ	34905	OM-MEX	Mar–Nov 2023	Lsd from AirExplore; rtl
Boeing 737-8H6	41767	G-JZDE	Jan 2024–present	Lsd from BOC; entered service May 2024
	41768	G-JZDF	Feb 2024–present	Lsd from BOC; entered service May 2024
	41769	G-JZDG	Mar 2024–present	Lsd from BOC; entered service July 2024

Type	C/n	Registration	Date	Notes/Fate
Boeing 737-8HX	36552	G-JZBX	Mar 2023–present	Lsd from ACG; entered service May 2023
Boeing 737-8JP	37818	G-DRTO	May 2019–present	Lsd from Tokyo Century Corporation; entered service Jun 2019
	40866	G-JZDD	Nov 2023–present	Lsd from Bank of America Leasing; entered service Mar 2024
Boeing 737-8K2	28375	G-GDFC	Jan 2011–present	Lsd from Volito Aviation until Oct 2015, Seraph Aviation Oct 2015–2022, Crestone Air Partners from 2022
	30390	G-JZHE	Mar 2016–present	
	28378	G-JZHF	Mar 2016–present	
Boeing 737-8K5	30882	9A-LAB	Mar–Nov 2024; Mar–Nov 2025	Lsd from ETF Airways
	28228	G-DRTE	Jul 2018–present	
	27982	G-GDFD	Apr 2011–present	
	30416	G-GDFU	Oct 2014–present	
	27986	G-GDFW	Mar 2014–present	Lsd from K2 Aircraft Leasing
	27987	G-GDFX	May 2014–present	Lsd from AerCap until Feb 2019, then Altavair AirFinance
	30417	G-JZHA	Jan 2015–present	
	28623	G-JZHB	Mar 2015–present	Lsd from AerCap until Dec 2017, then Aircastle
	30593	G-JZHC	Apr 2015–present	Lsd from AerCap until 2017, then Deucalion Aviation
Boeing 737-8KN	40260	G-DRTK	Mar 2022–present	Lsd from JP Lease; entered service May 2022
	40261	G-JZDB	Mar 2023–present	Lsd from SDH Wings; entered service Jul 2023

Type	C/n	Registration	Date	Notes/Fate
Boeing 737-8MG	63157	G-JZBA	Oct 2017–present	
	63158	G-JZBB	Oct 2017–present	
	63160	G-JZBC	Nov 2017–present	
	63159	G-JZBD	Nov 2017–present	
	63161	G-JZBE	Dec 2017–present	
	63162	G-JZBF	Dec 2017–present	
	63164	G-JZBG	Jan 2018–present	
	63163	G-JZBH	Jan 2018–present	
	63166	G-JZBI	Feb 2018–present	
	63165	G-JZBJ	Feb 2018–present	
	63167	G-JZBK	Mar 2018–present	
	63169	G-JZBL	Mar 2018–present	
	63170	G-JZBM	Apr 2018–present	
	63168	G-JZBN	Apr 2018–present	
	64439	G-JZBO	Aug 2018–present	
	64440	G-JZBP	Oct 2018–present	
	64441	G-JZBR	Nov 2018–present	
	64442	G-JZBS	Jan 2019–present	
	63144	G-JZHJ	Sep 2016–present	
	63145	G-JZHK	Oct 2016–present	
	63568	G-JZHL	Oct 2016–present	
	63570	G-JZHM	Nov 2016–present	
	63146	G-JZHN	Nov 2016–present	
	63569	G-JZHO	Nov 2016–present	
	63147	G-JZHP	Dec 2016–present	
	63148	G-JZHR	Dec 2016–present	
	63149	G-JZHS	Jan 2017–present	
	63150	G-JZHT	Jan 2017–present	
	63151	G-JZHU	Feb 2017–present	
	63152	G-JZHV	Feb 2017–present	
	63153	G-JZHW	Mar 2017–present	
	63154	G-JZHX	Mar 2017–present	
	63155	G-JZHY	Apr 2017–present	
	63156	G-JZHZ	Apr 2017–present	

Type	C/n	Registration	Date	Notes/Fate
Boeing 737-8Q8	30667	9A-ABC	Mar–Nov 2024; Mar–Nov 2025	Lsd from ETF Airways
	30722	G-DRTT	Dec 2018–present	Entered service Feb 2019
	35275	OK-TVH	Mar–Nov 2023	Lsd from Smartwings; rtl
	30688	OY-PSA	Apr–Oct 2016; May–Oct 2017	Lsd from Primera Air Scandinavia; rtl
Boeing 737-8S3	29250	OK-TSA	Apr 2013–Mar 2014	Lsd from Travel Service; rtl
Boeing 737-8Z9	28177	G-GDFP	Mar 2013–present	Entered service Apr 2013
	30421	G-GDFR	Mar 2013–present	
Boeing 757-200				
Boeing 757-204	26966	G-LSAM	Mar 2011–Oct 2012	Lsd from Allegiant Air; rtl
	26967	G-LSAL	Mar 2011–Oct 2012	Lsd from Allegiant Air; rtl
Boeing 757-21B	24014	G-LSAG	Nov 2006–Nov 2019	Wfu
	24015	G-LSAH	Nov 2006–Mar 2020	Wfu
	24016	G-LSAI	Nov 2006–Jan 2025	Wfu
Boeing 757-225	22689	G-LSAF	Oct 2006–Jan 2007	Lsd from GATX; rtl
Boeing 757-236	24122	G-LSAA	Oct 2005–Nov 2023	Opby Channel Express until Jan 2006; wfu
	24397	G-LSAD	Jun 2006–Mar 2020	Wfu
	24793	G-LSAJ	May 2008–Nov 2024	Wfu
Boeing 757-23A	25488	G-LSAC	May 2006–Nov 2024	Acquired by Dart Group in Mar 2006; wfu
Boeing 757-23N	27973	G-LSAK	May 2012–Jan 2025	Wfu
Boeing 757-256	26241	EC-ISY	May–Sep 2015	Lsd from Privilege Style; rtl
	29308	G-POWH	Mar–Oct 2016; May–Sep 2018	Lsd from Titan Airways; rtl

Type	C/n	Registration	Date	Notes/Fate
Boeing 757-27B	24136	G-LSAB	Dec 2005–Jan 2024	Opby Channel Express until Jan 2006; wfu
	24135	G-LSAE	Jun 2006–Dec 2024	Wfu
Boeing 757-2K2	26635	G-LSAN	May 2012–Nov 2024	Entered service Aug 2012; wfu
British Aerospace 146-300	E3205	G-FLTC	Apr–Nov 2006; Mar–May 2007	Lsd from Flightline; rtl

Bibliography

Books

Halford-MacLeod, Guy, *British Airlines Volume Three: 1964 to Deregulation*, The History Press (2010)

Woodley, Charles, *Flying To The Sun*, The History Press (2016)

Reports

Dart Group plc Annual Report and Accounts 1992–2020

Jet2 plc Annual Report and Accounts 2021–2025

Magazines

Various issues of:

Aircraft Illustrated, Ian Allan Publishing

Airliner World, Key Publishing Ltd

Jets Monthly, Kelsey Publishing/Key Publishing Ltd

Electronic Media

Wickstead, Maurice J, *Airlines of the British Isles since 1919*, Air-Britain (Historians) Ltd (2014)

Websites

www.Key.Aero

Civil Aviation Authority (C.A.A.) UK Airline Data (https://www.caa.co.uk/data-and-analysis/uk-aviation-market/airlines/uk-airline-data/)